SAFETY SYMBOLS

SAFETY SYMBOLS	HAZARD	EXAMPLES	PRECAUTION	REMEDY
DISPOSAL	Special disposal procedures need to be followed.	certain chemicals, living organisms	Do not dispose of these materials in the sink or trash can.	Dispose of wastes as directed by your teacher.
BIOLOGICAL	Organisms or other biological materials that might be harmful to humans	bacteria, fungi, blood, unpreserved tissues, plant materials	Avoid skin contact with these materials. Wear mask or gloves.	Notify your teacher if you suspect contact with material. Wash hands thoroughly.
EXTREME TEMPERATURE	Objects that can burn skin by being too cold or too hot	boiling liquids, hot plates, dry ice, liquid nitrogen	Use proper protection when handling.	Go to your teacher for first aid.
SHARP OBJECT	Use of tools or glassware that can easily puncture or slice skin	razor blades, pins, scalpels, pointed tools, dissecting probes, broken glass	Practice common-sense behavior and follow guidelines for use of the tool.	Go to your teacher for first aid.
FUME	Possible danger to respiratory tract from fumes	ammonia, acetone, nail polish remover, heated sulfur, moth balls	Make sure there is good ventilation. Never smell fumes directly. Wear a mask.	Leave foul area and notify your teacher immediately.
ELECTRICAL	Possible danger from electrical shock or burn	improper grounding, liquid spills, short circuits, exposed wires	Double-check setup with teacher. Check condition of wires and apparatus.	Do not attempt to fix electrical problems. Notify your teacher immediately.
IRRITANT	Substances that can irritate the skin or mucous membranes of the respiratory tract	pollen, moth balls, steel wool, fiberglass, potassium permanganate	Wear dust mask and gloves. Practice extra care when handling these materials.	Go to your teacher for first aid.
CHEMICAL	Chemicals that can react with and destroy tissue and other materials	bleaches such as hydrogen peroxide; acids such as sulfuric acid, hydrochloric acid; bases such as ammonia, sodium hydroxide	Wear goggles, gloves, and an apron.	Immediately flush the affected area with water and notify your teacher.
TOXIC	Substance may be poisonous if touched, inhaled, or swallowed	mercury, many metal compounds, iodine, poinsettia plant parts	Follow your teacher's instructions.	Always wash hands thoroughly after use. Go to your teacher for first aid.
OPEN FLAME	Open flame may ignite flammable chemicals, loose clothing, or hair	alcohol, kerosene, potassium permanganate, hair, clothing	Tie back hair. Avoid wearing loose clothing. Avoid open flames when using flammable chemicals. Be aware of locations of fire safety equipment.	Notify your teacher immediately. Use fire safety equipment if applicable.

Eye Safety
Proper eye protection should be worn at all times by anyone performing or observing science activities.

Clothing Protection
This symbol appears when substances could stain or burn clothing.

Animal Safety
This symbol appears when safety of animals and students must be ensured.

Radioactivity
This symbol appears when radioactive materials are used.

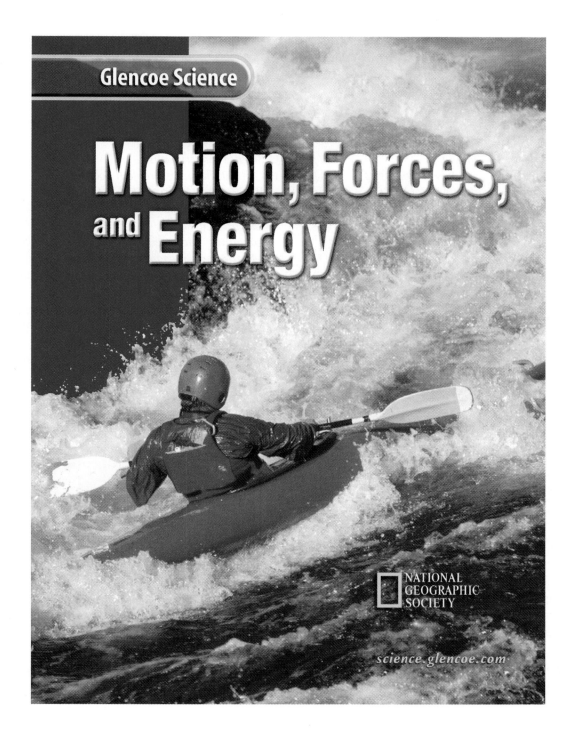

Glencoe Science

Motion, Forces, and Energy

NATIONAL GEOGRAPHIC SOCIETY

science.glencoe.com

Glencoe McGraw-Hill

New York, New York   Columbus, Ohio   Woodland Hills, California   Peoria, Illinois

Glencoe Science

Motion, Forces, and Energy

Student Edition
Teacher Wraparound Edition
Interactive Teacher Edition CD-ROM
Interactive Lesson Planner CD-ROM
Lesson Plans
Content Outline for Teaching
Dinah Zike's Teaching Science with Foldables
Directed Reading for Content Mastery
Foldables: Reading and Study Skills
Assessment
    Chapter Review
    Chapter Tests
    ExamView Pro Test Bank Software
    Assessment Transparencies
    Performance Assessment in the Science Classroom
    The Princeton Review Standardized Test Practice Booklet
Directed Reading for Content Mastery in Spanish
Spanish Resources
English/Spanish Guided Reading Audio Program
Reinforcement

Enrichment
Activity Worksheets
Section Focus Transparencies
Teaching Transparencies
Laboratory Activities
Science Inquiry Labs
Critical Thinking/Problem Solving
Reading and Writing Skill Activities
Mathematics Skill Activities
Cultural Diversity
Laboratory Management and Safety in the Science Classroom
MindJogger Videoquizzes and Teacher Guide
Interactive CD-ROM with Presentation Builder
Vocabulary PuzzleMaker Software
Cooperative Learning in the Science Classroom
Environmental Issues in the Science Classroom
Home and Community Involvement
Using the Internet in the Science Classroom

THE PRINCETON REVIEW

"Study Tip," "Test-Taking Tip," and "Test Practice" features in this book were written by The Princeton Review, the nation's leader in test preparation. Through its association with McGraw-Hill, The Princeton Review offers the best way to help students excel on standardized assessments.

The Princeton Review is not affiliated with Princeton University or Educational Testing Service.

Glencoe/McGraw-Hill

A Division of The McGraw-Hill Companies

Cover Images: A kayaker battles the whitewater rapids on the Thompson River in British Columbia, Canada.

Send all inquiries to:
Glencoe/McGraw-Hill
8787 Orion Place
Columbus, OH 43240

ISBN 0-07-825607-0
Printed in the United States of America.
3 4 5 6 7 8 9 10 027/111 06 05 04 03 02

Authors

National Geographic Society
Education Division
Washington, D.C.

Deborah Lillie
Math and Science Writer
Sudbury, Massachusetts

Thomas McCarthy, PhD
Science Department Chair
St. Edwards School
Vero Beach, Florida

Cathy Ezrailson, PhD
Science Department Head
Academy for Science and Health Professions
Conroe, Texas

Dinah Zike
Educational Consultant
Dinah-Might Activities, Inc.
San Antonio, Texas

Margaret K. Zorn
Science Writer
Yorktown, Virginia

Content Consultants

Alan Bross, PhD
High-Energy Physicist
Fermilab
Batavia, Illinois

Jack Cooper
Adjunct Faculty Math and Science
Navarro College
Corsicana, Texas

Sandra K. Enger, PhD
Coordinator
UAH, Institute for Science Education
Huntsville, Alabama

Carl Zorn, PhD
Staff Scientist
Jefferson Laboratory
Newport News, Virginia

Series Safety Consultants

Aileen Duc, PhD
Science II Teacher
Hendrick Middle School
Plano, Texas

Malcolm Cheney, PhD
OSHA Chemical Safety
Officer
Hall High School
West Hartford, Connecticut

Sandra West, PhD
Associate Professor of
Biology
Southwest Texas State
University

Series Math Consultant

Michael Hopper, DEng
Manager of Aircraft Certification
Raytheon Company
Greenville, Texas

Teri Willard, EdD
Department of Mathematics
Montana State University
Belgrade, Montana

CONTENTS

Thermal Energy—156

Field Guides

Skill Handbooks—194

Reference Handbook

English Glossary—223

Spanish Glossary—227

Index—232

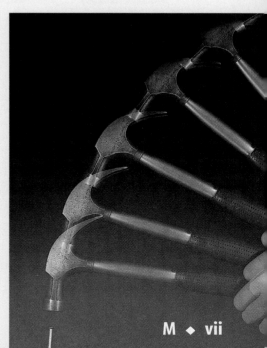

Interdisciplinary Connections/Activities

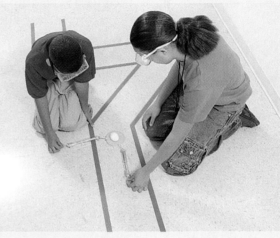

Feature Contents

Activities/Science Connections

EXPLORE ACTIVITY

Problem-Solving Activities

Math Skills Activities

Skill Builder Activities

Science

Math

Technology

Science
INTEGRATION

SCIENCE Online

THE PRINCETON REVIEW

Feature Contents

Science in Motion

How do scientists learn more about the world? Scientists usually follow an organized set of procedures to solve problems. These procedures are called scientific methods. Although the steps in these methods can vary depending on the type of problem a scientist is solving, they all are an organized way of asking a question, forming a possible answer, investigating the answer, and drawing conclusions about the answer. Humans have investigated questions about motion for thousands of years, asking questions such as: "What causes motion? How fast do things fall? How does a pendulum work?" However, scientific methods were not always used to learn the answers to these questions.

Figure 1
Motion, like on this busy road, is all around you.

Early Scientists

The ancient Greeks believed in supernatural beings—gods and goddesses—whose powers made the world work. In the 500s B.C., a group of Greek philosophers in the city of Miletus proposed that natural events should only be explained by what humans can learn with their senses—sight, hearing, smell, touch, and taste. 200 years later, Aristotle, a Greek philosopher and teacher, developed a system of logic for distinguishing truth from falsehood. He also studied plants and animals and recorded detailed observations of them. Because of these practices, he is considered one of the first scientists as they are defined today.

Figure 2
At his school in ancient Athens, Greece, Aristotle taught philosophy and science.

Aristotle also investigated how objects move and why they continue moving. He believed that the speed of a falling object depends on its weight. Unfortunately, Aristotle did not have a scientific method to test his ideas. Therefore, Aristotle's untested theory was not proven wrong for hundreds of years. In the late 1500s, Italian scientist Galileo Galilei conducted experiments to test Aristotle's ideas. He rolled balls down inclined planes and swung pendulums, measuring how far and how fast they moved. According to legend, Galileo climbed to the top of the Tower of Pisa and dropped two objects of different weights. They hit the ground at the same time, finally proving that the speed of a falling object doesn't depend on its weight.

Figure 3
Galileo made observations of the motion of pendulums in order to learn about motion.

Developing Scientific Methods

Galileo and others that came after him gradually developed new ideas about how to learn about the universe. These new methods of scientific investigation were different from the methods used by earlier philosophers in an important way. A scientific explanation makes predictions that can be tested by observations of the world or by doing experiments. If the predictions are not supported by the observations or experiments, the scientific explanation cannot be true and has to be changed or discarded.

Figure 4
A scientific explanation of motion would include the motion of this roller coaster.

Physical Science

The study of motion, forces, and energy is part of physical science. Physical scientists also learn about chemicals, atoms, electricity, sound, and more.

Like all scientists, they use experimentation and careful observation to answer questions about how the world works. Other scientists learn about these experiments and try to repeat them. In this way, scientists eliminate the flaws in their work and participate in the search for answers.

Scientific Methods

1. Identify a question.
Determine a question to be answered.

2. Form a hypothesis.
Gather information and make an educated guess about the answer to the question.

3. Test the hypothesis.
Perform experiments or make observations to see if the hypothesis is supported.

4. Analyze results.
Look for patterns in the data that have been collected.

5. Draw a conclusion.
Decide what the test results mean. Communicate your results.

Scientists use scientific methods to answer questions about the forces of motion.

Scientific Methods

The understanding of motion was undertaken by philosophers such as Descartes and scientists such as Galileo. Their efforts led to the creation of procedures, called scientific methods, which scientists use to investigate the world. Scientific methods generally include several steps.

Identifying a Question

The first step in a scientific method is to identify a question to be answered. For example, Aristotle wanted to know what causes motion. The answer to one question often leads to others. Aristotle wondered how an object's weight affects the speed at which it falls. After Galileo showed that an object's weight does not affect its falling speed, Newton wanted to know how fast objects fall, regardless of their weight.

Forming a Hypothesis

The next step is to gather information and form a hypothesis, an educated guess about what the answer is. It is useful to limit the information used to form a hypothesis. For example, the Greeks of Miletus decided that religious beliefs were not an acceptable basis for a hypothesis, but the five senses were acceptable. Aristotle's hypothesis—heavier objects fall faster—could have been based on observing that feathers seem to fall more slowly than rocks.

Testing a Hypothesis

A hypothesis must be testable to see if it is correct. This is done by performing experiments and measuring the results. Galileo tested Aristotle's hypothesis by rolling balls of differing weight down an inclined plane to see which, if either, rolled faster. Since a well-designed experiment is crucial, Galileo made sure the inclined plane was smooth and the balls were released in the same way.

Analyzing Results

Scientists collect information, called data, which must be analyzed. In order to organize, study, and detect patterns in data, scientists use graphs and other methods.

Collecting data requires careful measurements. Many experiments of the past were flawed because the measuring devices were inaccurate. Because Galileo needed precise timing, he used a water clock to measure the time for a ball to roll down the inclined plane. If his clock had been inaccurate, Galileo's results would have been less useful.

Drawing a Conclusion

The last step in a scientific experiment is to draw a conclusion based on results and observations. Sometimes the data does not support the original hypothesis and scientists must start the process again, beginning with a new hypothesis. Other times, though, the data supports the original hypothesis. If a hypothesis is supported by repeated experiments, it can become a theory—an idea that has withstood repeated testing and is used to explain observations. Scientists, however, know that nothing is certain. A new idea, a new hypothesis, and a new experiment can alter what is believed to be true about the world.

Figure 5
These students are conducting an experiment to learn how objects move.

You Do It

A ball may fall, but will it bounce back? What determines how high and how fast it will bounce? Make a list of possible factors that affect the way a ball bounces. Choose one of these and form a hypothesis about it. Think of experiments you could do to test your hypothesis.

Motion and Momentum

Racers, like the ones shown here, want to know who is the fastest. How can you determine who's the fastest? What has to be measured to determine a racer's speed? How can you describe motion when speed is changing? In this chapter you will learn how to describe motion, including motion that is changing. You will also study how motion changes when objects collide.

What do you think?

Science Journal Look at the picture below with a classmate. Discuss what this might be or what is happening. Here's a hint: *You can see these almost everywhere, day or night.* Write your answer or your best guess in your Science Journal.

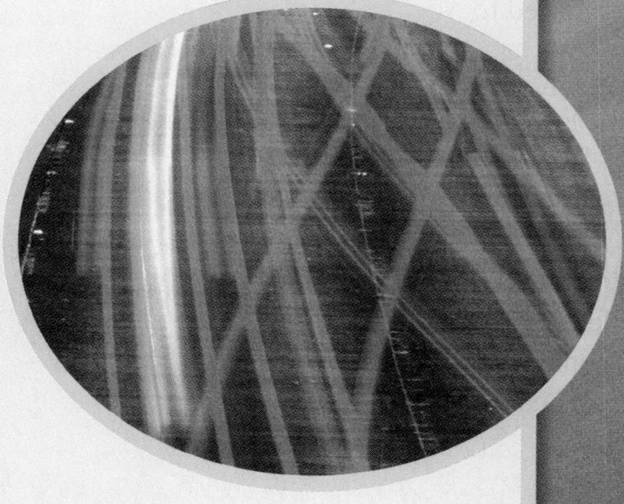

EXPLORE ACTIVITY

How is it possible for a 70-kg football player to knock down a 110-kg football player? The smaller player usually must be running faster. Mass makes a difference when two objects collide, but the speed of the objects also matters. Explore the behavior of colliding objects during this activity.

Model collisions

1. Space yourself about 2 m away from a partner. Slowly roll a baseball on the floor toward your partner, and have your partner roll a baseball quickly into your ball.

2. Have your partner slowly roll a baseball as you quickly roll a tennis ball into the baseball.

3. You and your partner roll two tennis balls toward each other at the same speed.

Observe

Describe your observations of each of these collisions. In your Science Journal, write a paragraph discussing how the motion of the balls changed after the collision.

Before You Read

FOLDABLES
Reading & Study Skills

Making a Vocabulary Study Fold Knowing the definition of vocabulary words is a good way to ensure that you understand the content of the chapter.

1. Place a sheet of notebook paper in front of you so the short side is at the top and the holes are on the right side. Fold the paper in half from the left side to the right side.

2. Through the top thickness of paper, cut along every third line from the outside edge to the fold, forming tabs.

3. Before you read, write the vocabulary words from each section in this chapter on the front of the tabs. Under each tab, write what you think the word means.

4. As you read the chapter, add to and correct your definitions.

1 What is motion?

Research Visit the Glencoe Science Web site at **science.glencoe.com** for more information about early attempts to study motion. Make a table to show what you learn.

Matter and Motion

All matter in the universe is constantly in motion, from the revolution of Earth around the Sun to electrons moving around the nucleus of an atom. Plants grow. Lava flows from a volcano. Bees move from flower to flower as they gather pollen. Blood circulates through your body. These are all examples of matter in motion. How can the motion of these different objects be described?

Changing Position

To describe an object in motion, you must recognize first that the object is in motion. Something is in motion if it is changing position. It could be a fast-moving airplane, a leaf swirling in the wind, or water trickling from a hose. Even your school is moving through space attached to Earth. When an object moves from one location to another, it is changing position. The runners shown in **Figure 1** sprint from the start line to the finish line. Their positions change so they are in motion.

Figure 1
When running a race, you are in motion because your position changes.

Relative Motion Determining whether something changes position requires a point of reference. An object changes position if it moves relative to a reference point. To visualize this, picture yourself competing in a 100-m dash. You begin just behind the start line. When you pass the finish line, you are 100 m from the start line. If the start line is your reference point then your position has changed by 100 m relative to the start line, and motion has occurred. Look at **Figure 2.** How can you determine that the dog has been in motion?

✔ Reading Check *How do you know if an object has changed position?*

Distance and Displacement Suppose you are to meet your friends at the park in five minutes. Can you get there on time by walking, or should you ride your bike? To help you decide, you need to know the distance you will travel to get to the park. This distance is the length of the route you will travel from your house to the park.

Suppose the distance you traveled from your house to the park was 200 m. When you get to the park, how would you describe your location? You could say that your location was 200 m from your house. To describe your location exactly, you also would have to tell in what direction you traveled. Did you travel 200 m east or 200 m west? Your final position would depend on the distance traveled and the direction. To describe your location, you would specify your displacement. Displacement includes the distance between the starting and stopping points, and the direction in which you travel. **Figure 3** shows the difference between distance and displacement.

Figure 2
Motion occurs when something moves relative to a reference point. *The dog has moved relative to what object?*

Figure 3
Distance is how far you have walked. Displacement is the direction and difference in position between your starting point and your ending point.

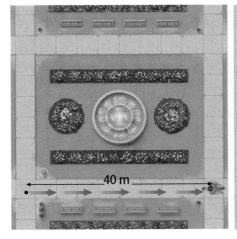

Distance: 40 m
Displacement: 40 m east

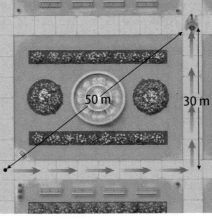

Distance: 70 m
Displacement: 50 m northeast

Distance: 140 m
Displacement: 0 m

Speed

Knowing how fast something is moving can be important. The faster something is moving, the less time it takes to travel a certain distance. **Speed** is the distance traveled divided by the time taken to travel the distance. This definition can be written as the following equation:

$$\text{speed} = \frac{\text{distance}}{\text{time}}$$

For example, the fastest runners can run the 100-m dash in about 10 s. When sprinters run 100 m in 10 s, their speed is as follows:

$$\text{speed} = \frac{\text{distance}}{\text{time}}$$

$$= \frac{100 \text{ m}}{10 \text{ s}}$$

$$= 10 \text{ m/s}$$

The units of speed are units of distance divided by units of time. In SI units, the units of speed are meters per second (m/s).

Life Science INTEGRATION

Different animals can move at different top speeds. What are some of the fastest animals? Research the characteristics that help animals run, swim, or fly at high speed.

Math Skills Activity

Calculating Speed

Example Problem

Calculate the speed of a swimmer who swims 100 m in 56 s.

Solution

1 *This is what you know:* distance: 100 m
time: 56 s

2 *This is what you need to know:* speed

3 *This is the equation you need to use:* speed = distance/time

4 *Substitute the known values:* speed = (100 m)/(56 s)
speed = 1.8 m/s

Check your answer by multiplying the calculated speed by the time. Did you calculate the distance that was given in the problem?

Practice Problem

A runner completes a 400-m race in 43.9 s. In a 100-m race, he finishes in 10.4 s. In which race was his speed faster?

For more help, refer to the Math Skill Handbook.

Average Speed If a sprinter ran the 100-m dash in 10 s, she probably couldn't have run the entire race with a speed of 10 m/s. Consider that when the race started, the sprinter wasn't moving. Then, as she started running, she moved faster and faster, which increased her speed. During the entire race, the sprinter's speed could have been different from instant to instant. However, the sprinter's motion for the entire race can be described by her average speed, which is 10 m/s. **Average speed** is found by dividing the total distance traveled by the time taken.

✔ Reading Check *How is average speed calculated?*

An object in motion can change speeds many times as it speeds up or slows down. The speed of an object at one instant of time is the object's **instantaneous speed.** To understand the difference between average and instantaneous speeds, think about walking to the library. If it takes you 0.5 h to walk 2 km to the library, your average speed would be as follows:

$$\text{speed} = \frac{\text{distance}}{\text{time}}$$

$$= \frac{2 \text{ km}}{0.5 \text{ h}} = 4 \text{ km/h}$$

However, you might not have been moving at the same speed throughout the trip. At a crosswalk, your instantaneous speed might have been 0 km/h. If you raced across the street, your speed might have been 7 km/h. If you were able to walk at a steady rate of 4 km/h during the entire trip, you would have moved at a constant speed. Average speed, instantaneous speed, and constant speed are illustrated in **Figure 4.**

Measuring Average Speed

Procedure
1. Measure the distance between two marks, such as two doorways.
2. Time yourself walking from one mark to the other.
3. Time yourself walking slowly, walking safely and quickly, and walking with a varying speed; for example, slow/fast/slow.

Analysis
1. Calculate your average speed in each case.
2. Predict how long it would take you to walk 100 m slowly, at your normal speed, and quickly.

Figure 4
The average speed of each ball is the same from 0 s to 4 s.

A This ball is moving at a constant speed. In each second, the ball moves the same distance.

B This ball has a varying speed. Its instantaneous speed is fast between 0 s and 1 s and slow between 2 s and 3 s.

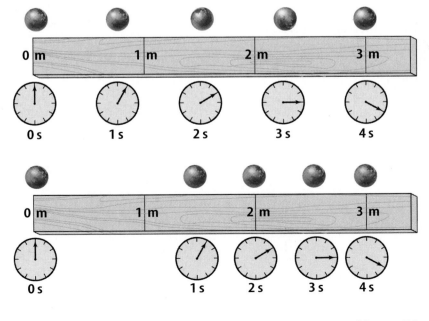

Graphing Motion

You can represent the motion of an object with a distance-time graph. For this type of graph, time is plotted on the horizontal axis and distance is plotted on the vertical axis. **Figure 5** shows the motion of two students who walked across a classroom, plotted on a distance-time graph.

Distance-Time Graphs and Speed The distance-time graph can be used to compare the speeds of objects. Look at the graph shown in **Figure 5.** According to the graph, after 1 s student A traveled 1 m. Her average speed during the first second is as follows:

$$\text{speed} = \frac{\text{distance}}{\text{time}} = \frac{1 \text{ m}}{1 \text{ s}} = 1 \text{ m/s}$$

Student B, however, only traveled 0.5 m in the first second. His average speed is

$$\text{speed} = \frac{\text{distance}}{\text{time}} = \frac{0.5 \text{ m}}{1 \text{ s}} = 0.5 \text{ m/s}$$

So student A traveled faster than student B. Now compare the steepness of the lines on the graph in **Figure 5.** The line representing the motion of student A is steeper than the line of student B. A steeper line on the distance-time graph represents a greater speed. A horizontal line on the distance-time graph means that no change in position occurs. Then the speed, represented by the line on the graph, is zero.

Figure 5
The motion of two students walking across a classroom is plotted on this distance-time graph.
Which student moved faster?

Distance versus Time

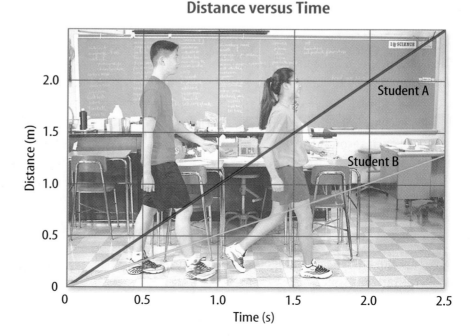

Velocity

If you are hiking in the woods, it is important to know in which direction you should walk in order to get back to camp. You want to know not only your speed, but also the direction in which you are moving. The **velocity** of an object is the speed of the object and direction of its motion. This is why a compass and a map, like the one shown in **Figure 6,** are useful to hikers. To get back to camp before nightfall, they need to know how far, how fast, and in what direction they need to travel. The map and the compass help the hikers to determine what their velocity must be. Velocity has the same units as speed, but it also includes the direction of motion.

The velocity of an object can change if the object's speed changes, its direction of motion changes, or they both change. For example, suppose a car is traveling at a speed of 60 km/h north and then turns left at an intersection and continues on with a speed of 60 km/h. The speed of the car is constant at 60 km/h, but the velocity changes from 60 km/h north to 60 km/h west. Why can you say the velocity of a car changes as it comes to a stop at an intersection?

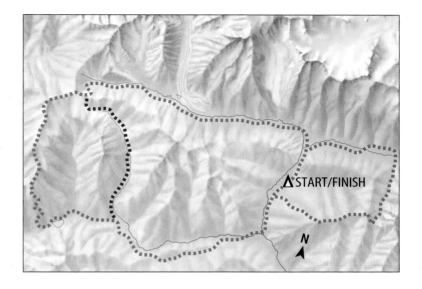

Figure 6
A map helps determine the direction in which you need to travel.

Section 1 Assessment

1. A dancer moves 5 m toward the left of the stage over the course of 15 s. What is her average velocity for this time?

2. If you know an object's velocity, do you know its speed? Explain.

3. An airplane flies a distance of 650 km at an average speed of 300 km/h. How much time did the flight take?

4. **Think Critically** A bee flies 25 m north of the hive, then 10 m east, 5 m west, and 10 m south. How far north and east of the hive is it now? Explain how you calculated your answer.

Skill Builder Activities

5. **Making and Using Graphs** You walk forward at 1.5 m/s for 8 s. Your friend decides to walk faster and starts out at 2.0 m/s for the first 4 s. But then she slows down and walks forward at 1.0 m/s for the next 4 s. Make a distance-time graph of your motion and your friend's motion. Who walked farther? **For more help, refer to the** Science Skill Handbook.

6. **Using a Database** Use a database to research the top speeds of different animals. Convert all data to units of m/s. **For more help, refer to the** Technology Skill Handbook.

Acceleration

As You Read

What **You'll Learn**

■ **Define** acceleration.
■ **Predict** what effect acceleration will have on motion.

Vocabulary
acceleration

Why **It's Important**
Whenever an object changes its motion, it accelerates.

Acceleration and Motion

When you watch the first few seconds of a liftoff, a rocket barely seems to move. With each passing second, however, you can see it move faster until it reaches an enormous speed. How could you describe the change in the rocket's motion? When an object changes its motion, it is accelerating. **Acceleration** is the change in velocity divided by the time it takes for the change to occur.

Like velocity, acceleration has a direction. If an object speeds up, the acceleration is in the direction that the object is moving. If an object slows down, the acceleration is opposite to the direction that the object is moving. What if the direction of the acceleration is at an angle to the direction of motion? Then the direction of motion will turn toward the direction of the acceleration.

Speeding Up You get on a bicycle and begin to pedal. The bike moves slowly at first, then it accelerates because its speed increases. When an object that is already in motion speeds up, it also is accelerating. Imagine that you are biking along a level path and you start pedaling harder. Your speed increases. When its speed is increasing, an object is accelerating.

Suppose a toy car is speeding up, as shown in **Figure 7.** Each second, the car moves at a greater speed and travels a greater distance than it did in the previous second. When the car stops accelerating, it will move in a straight line at the speed it reached when the acceleration stopped.

Figure 7
The toy car is accelerating to the right. The speed is increasing.

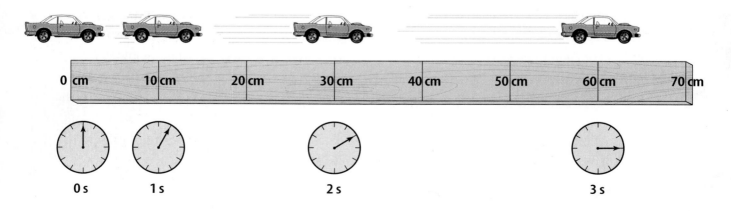

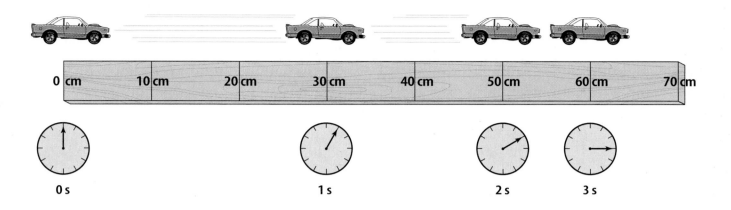

Slowing Down Now suppose you are biking at a speed of 4 m/s and you apply the brakes. This causes you to slow down. It might sound odd, but because your speed changes, you have accelerated. Acceleration occurs when an object slows down, as well as when it speeds up. The car in **Figure 8** is slowing down. During each time interval, the car travels a smaller distance, so its speed is decreasing.

In each of these examples, speed is changing, so acceleration is occurring. Because speed is decreasing, the direction of the acceleration is opposite to the direction of motion. Any time an object slows down, its acceleration is in the opposite direction of its motion.

Changing Direction Motion is not always along a straight line. If the acceleration is at an angle to the direction of motion, the object will turn. At the same time, it might speed up, slow down, or have no change in speed.

Picture yourself again riding a bicycle. When you lean to one side and turn the handlebars, the bike turns. Because the direction of the bike's motion has changed, the bike has accelerated. The acceleration is in the direction that the bicycle turned.

Figure 9 shows another example of an object that is accelerating. The ball starts moving upward, but its direction of motion changes as its path turns downward. Here the acceleration is downward. The longer the ball accelerates, the more its path turns toward the direction of acceleration.

Figure 8
The car is moving to the right but accelerating to the left. In each time interval, it covers less distance and moves more slowly.

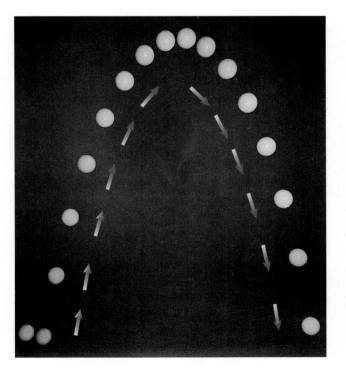

Figure 9
The ball starts out by moving forward and upward, but the acceleration is downward, so the ball's path turns in that direction.

Calculating Acceleration

If an object is moving in a straight line, its acceleration can be calculated using this equation.

$$\text{acceleration} = \frac{\text{final speed} - \text{initial speed}}{\text{time}}$$

In this equation, the final speed is the speed at the end of the time period and the initial speed is the speed at the beginning of the time period. Also, time is the length of time over which the motion changes.

This equation also can be written in a simpler way by using symbols. Let a stand for acceleration and t stand for time. Then let s_f stand for the final speed and s_i stand for the initial speed. Then the above equation can be written as follows.

$$a = \frac{(s_f - s_i)}{t}$$

The unit of acceleration is distance divided by time squared. In SI units, acceleration has units of meters per second squared (m/s^2).

Math Skills Activity

Calculating Acceleration

Example Problem

Calculate the acceleration of a bus whose speed changes from 6 m/s to 12 m/s over a period of 3 s.

1 *This is what you know:*   initial speed: s_i = 6 m/s
final speed: s_f = 12 m/s
time: t = 3 s

2 *This is what you need to know:*   acceleration: a

3 *This is the equation you need to use:* $a = (s_f - s_i)/t$

4 *Substitute the known values:* $a = (12\text{ m/s} - 6\text{ m/s})/(3\text{ s}) = (6\text{ m/s})/(3\text{ s})$
$= 2\text{ m/s}^2$

Check your answer by multiplying the calculated acceleration by the time. Then add the initial speed. Did you calculate the final speed given in the problem?

Practice Problem

A train's velocity increases from 7 m/s to 18 m/s over a period of 120 s. Calculate its acceleration.

For more help, refer to the Math Skill Handbook.

Figure 10
When skidding to a stop, you are slowing down. This means you have a negative acceleration.

Positive and Negative Acceleration An object is accelerating when it speeds up, and the acceleration is in the same direction as the motion. An object also is accelerating when it slows down, but the acceleration is in the direction opposite the motion, such as the bicycle in **Figure 10.** How else is acceleration different when an object is speeding up and slowing down?

Suppose you were riding your bicycle in a straight line and speeded up from 4 m/s to 6 m/s in 5 s. You could calculate your acceleration from the equation on the previous page.

$$a = \frac{(s_f - s_i)}{t}$$
$$= \frac{(6 \text{ m/s} - 4 \text{ m/s})}{5 \text{ s}} = \frac{+2 \text{ m/s}}{5 \text{ s}}$$
$$= +0.4 \text{ m/s}^2$$

When you speed up, your final speed always will be greater than your initial speed. So subtracting the initial speed from the final speed gives a positive number. As a result, your acceleration is positive when you are speeding up.

Suppose you slow down from a speed of 4 m/s to 2 m/s in 5 s. Now the final speed is less than the initial speed. You could calculate your acceleration as follows:

$$a = \frac{(s_f - s_i)}{t}$$
$$= \frac{(2 \text{ m/s} - 4 \text{ m/s})}{5 \text{ s}} = \frac{-2 \text{ m/s}}{5 \text{ s}}$$
$$= -0.4 \text{ m/s}^2$$

Because your final speed is less than your initial speed, your acceleration is negative when you slow down.

Modeling Acceleration

Procedure
1. Use **masking tape** to lay a course on the floor. Mark a starting point and place marks along a straight path at 10 cm, 40 cm, 90 cm, 160 cm, and 250 cm from the start.
2. Clap a steady beat. On the first beat, the person walking the course is at the starting point. On the second beat, the walker is on the first mark, and so on.

Analysis
1. Describe what happens to your speed as you move along the course. Infer what would happen if the course were extended farther.
2. Repeat step 2, starting at the other end. Are you still accelerating? Explain.

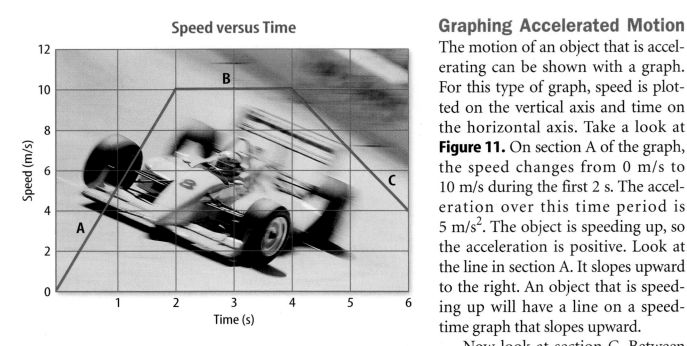

Speed versus Time

Graphing Accelerated Motion

The motion of an object that is accelerating can be shown with a graph. For this type of graph, speed is plotted on the vertical axis and time on the horizontal axis. Take a look at **Figure 11.** On section A of the graph, the speed changes from 0 m/s to 10 m/s during the first 2 s. The acceleration over this time period is 5 m/s^2. The object is speeding up, so the acceleration is positive. Look at the line in section A. It slopes upward to the right. An object that is speeding up will have a line on a speed-time graph that slopes upward.

Now look at section C. Between 4 s and 6 s the speed changes from 10 m/s to 4 m/s. The acceleration is −3 m/s^2. The object is slowing down, so the acceleration is negative. On the speed-time graph, the line in section C is sloping downward to the right. An object that is slowing down will have a line on a speed-time graph that slopes downward.

On section B, where the line is horizontal, the change in speed is zero. So a horizontal line on the speed-time graph represents an acceleration of zero or constant speed.

Figure 11
The speed-time graph can be used to find acceleration. When the line rises, the object is speeding up. When it is horizontal, the acceleration is zero. When the line falls, the object is slowing down.

✔ **Reading Check** *How is an acceleration of zero represented on a speed-time graph?*

Section 2 Assessment

1. A runner accelerates from 0 m/s to 3 m/s in 12 s. What was the acceleration?
2. A speed-time graph shows a line sloping downward. How was the speed changing?
3. In what three ways can acceleration change an object's motion?
4. An object falls with an acceleration of 9.8 m/s^2. What is its speed after 2 s?
5. **Think Critically** You start to roll backward down a hill on your bike, so you use the brakes to stop your motion. In what direction did you accelerate?

Skill Builder Activities

6. **Forming Operational Definitions** Give an operational definition of acceleration. **For more help, refer to the** Science Skill Handbook.
7. **Making and Using Graphs** A sprinter had the following speeds at different times during a race: 0 m/s at 0 s, 4 m/s at 2 s, 7 m/s at 4 s, 10 m/s at 6 s, 12 m/s at 8 s, and 10 m/s at 10 s. Plot these data on a speed-time graph. During what time intervals is the acceleration positive? Negative? Is the acceleration ever zero? **For more help, refer to the** Science Skill Handbook.

3 Momentum

Mass and Inertia

The world you live in is filled with objects in motion. How can you describe these objects? Objects have many properties such as color, size, and composition. One important property of an object is its mass. The **mass** of an object is the amount of matter in the object. In SI units the unit for mass is the kilogram.

The weight of an object is related to the object's mass. Objects with more mass weigh more than objects that have less mass. A bowling ball has more mass than a pillow, so it weighs more than a pillow. However, the size of an object is not the same as the mass of the object. For example, a pillow is larger than a bowling ball, but the bowling ball has more mass.

Objects with different masses are different in an important way. Think about what happens when you try to stop someone who is rushing toward you. A small child is easy to stop. A large adult is hard to stop. The more mass an object has, the harder it is to slow it down, speed it up, or turn it. This tendency of an object to resist a change in its motion is called **inertia.** Objects with more mass have more inertia, as shown in **Figure 12.** The more mass an object has, the harder it is to change its motion.

As You Read

What You'll Learn
- **Explain** the difference between mass and inertia.
- **Define** momentum.
- **Predict** motion using the law of conservation of momentum.

Vocabulary
mass
inertia
momentum
law of conservation of momentum

Why It's Important
Objects in motion have momentum. The motion of objects after they collide depends on their momentum.

Figure 12
The more mass an object has, the greater its inertia is. A table-tennis ball responds to a gentle hit that would move a tennis ball only slightly.

Momentum

You know that the faster a bicycle moves, the harder it is to stop. Just as increasing the mass of an object makes it harder to stop, increasing the speed or velocity of an object also makes it harder to stop. The **momentum** of an object is a measure of how hard it is to stop the object. This depends on the object's mass and velocity. The momentum of an object can be calculated from this equation.

$$\text{momentum} = \text{mass} \times \text{velocity}$$

Momentum is usually symbolized by p. If m stands for mass and v stands for velocity, this equation can be written like this.

$$p = mv$$

According to this equation, the momentum increases if the mass of the object or its velocity increases. Mass is measured in kilograms and velocity has units of meters per second, so momentum has units of kilograms multiplied by meters per second (kg·m/s). Also, because velocity includes a direction, momentum has a direction that is the same direction as its velocity.

Life Science

INTEGRATION

A running animal has momentum. A small animal might be able to turn more quickly than a larger pursuing predator, because the smaller animal has less momentum. The larger an animal is and the faster it runs, the harder it is to turn or stop. Research the sizes of some predators and their usual prey.

Math Skills Activity

Calculating Momentum

Example Problem

Calculate the momentum of a 14 kg bicycle traveling north at 2 m/s.

Solution

1️⃣ *This is what you know:*   mass: $m = 14\,\text{kg}$
velocity: $v = 2\,\text{m/s north}$

2️⃣ *This is what you need to find:*   momentum: p

3️⃣ *This is the equation you need to use:* $p = mv$

4️⃣ *Substitute the known values:* $p = 14\,\text{kg} \times 2\,\text{m/s north}$
     $= 28\,\text{kg·m/s north}$

Check your answer by dividing your momentum calculation by the mass of the bicycle. Did you calculate the velocity given in the problem?

Practice Problem

A 10,000-kg train is traveling east at 15 m/s. Calculate the momentum of the train.

For more help, refer to the Math Skill Handbook.

Conservation of Momentum

If you've ever played billiards, you know that when the cue ball hits another ball, the motion of both balls changes. The cue ball slows down and may change direction, so its momentum decreases. Meanwhile, the other ball starts moving, so its momentum increases. It seems as if momentum is transferred from the cue ball to the other ball.

In fact, during the collision the momentum lost by the cue ball was gained by the other ball. This means that the total momentum of both balls was the same just before and just after the collision. This is true for any collision, as long as no outside forces such as friction act on the objects and change their speeds after the collision. According to the **law of conservation of momentum,** the total momentum of objects that collide with each other is the same before and after the collision. This is true for the collisions of the billiard balls shown in **Figure 13,** as well as for atoms, cars, football players, or any other matter.

Using Momentum Conservation

Outside forces such as gravity and friction are almost always acting on objects that are colliding. However, sometimes the effects of these forces are small enough that they can be ignored. Then the law of conservation of momentum enables you to predict how the motions of objects will change after a collision.

There are many ways that collisions can occur. Sometimes the objects that collide will bounce off of each other, like the bowling ball and bowling pins in **Figure 14A.** In some collisions, objects will stick to each other after the collision, like the two football players in **Figure 14B.** In this type of collision, the law of conservation of momentum enables the speeds of the objects after the collision to be calculated.

Figure 13
When the cue ball hits the other billiard balls, it slows down because it transfers part of its momentum to the other billiard balls. *What would happen to the speed of the cue ball if all of its momentum were transferred to the other billiard balls?*

Figure 14
In these collisions, the total momentum before the collision equals the total momentum after the collision.

A When the bowling ball hits the pins, some of its momentum is transferred to the pins. The ball slows down and the pins speed up.

B When one player tackles the other, they both change speeds but momentum is conserved.

Figure 15
Momentum is conserved in the collision of the backpack and the student. **A** Before the student on skates and the backpack collide, she is not moving. **B** After the collision, the student and the backpack move together at a slower speed than the backpack had before the collision.

Sticking Together Picture yourself standing on a pair of skates when someone throws a backpack to you, as in **Figure 15.** When you catch the backpack, you and the backpack continue to move in the same direction that the backpack was moving before the collision.

The law of conservation of momentum can be used to find your speed or velocity after you catch the backpack. Suppose a backpack with a mass of 2 kg is tossed at a speed of 5 m/s. You have a mass of 48 kg, and initially you are at rest. Then the total momentum before the collision would be

$$\text{total momentum} = \text{momentum of backpack} + \text{your momentum}$$
$$= 2 \text{ kg} \times 5 \text{ m/s} + 48 \text{ kg} \times 0 \text{ m/s}$$
$$= 10 \text{ kg·m/s}$$

After the collision, the total momentum remains the same and only one object is moving. It has a combined mass of you and the backpack. You can use the equation for momentum to find the new velocity.

$$\text{total momentum} = (\text{mass of backpack} + \text{your mass}) \times \text{velocity}$$
$$10 \text{ kg·m/s} = (2 \text{ kg} + 48 \text{ kg}) \times \text{velocity}$$
$$10 \text{ kg·m/s} = (50 \text{ kg}) \times \text{velocity}$$
$$0.2 \text{ m/s} = \text{velocity}$$

This is your velocity right after you catch the backpack. The final velocity is much less than the initial velocity of the backpack. The velocity decreases because the mass of you and the backpack together is much greater than the mass of the backpack alone.

As you continue to move on your skates, the force of friction between the ground and the skates slows you down. As a result, the momentum of you and the backpack together continually decreases until you come to a stop.

NATIONAL GEOGRAPHIC

VISUALIZING CONSERVATION OF MOMENTUM

Figure 16

The law of conservation of momentum can be used to predict the results of collisions between different objects, whether they are subatomic particles smashing into each other at enormous speeds, or the collisions of marbles, as shown on this page. What happens when one marble hits another marble initially at rest? The results of the collisions depend on the masses of the marbles.

A Here, a less-massive marble strikes a more-massive marble that is at rest. After the collision, the smaller marble bounces off in the opposite direction. The larger marble moves in the same direction that the small marble was initially moving.

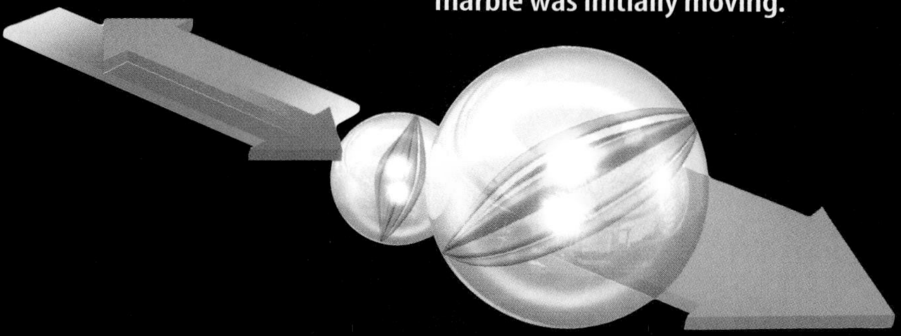

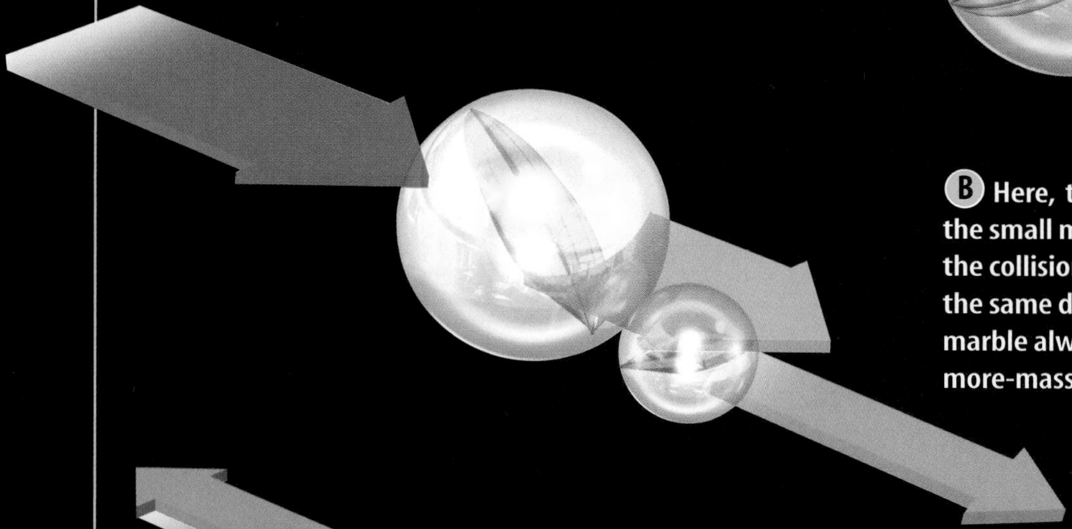

B Here, the large marble strikes the small marble that is at rest. After the collision, both marbles move in the same direction. The less-massive marble always moves faster than the more-massive one.

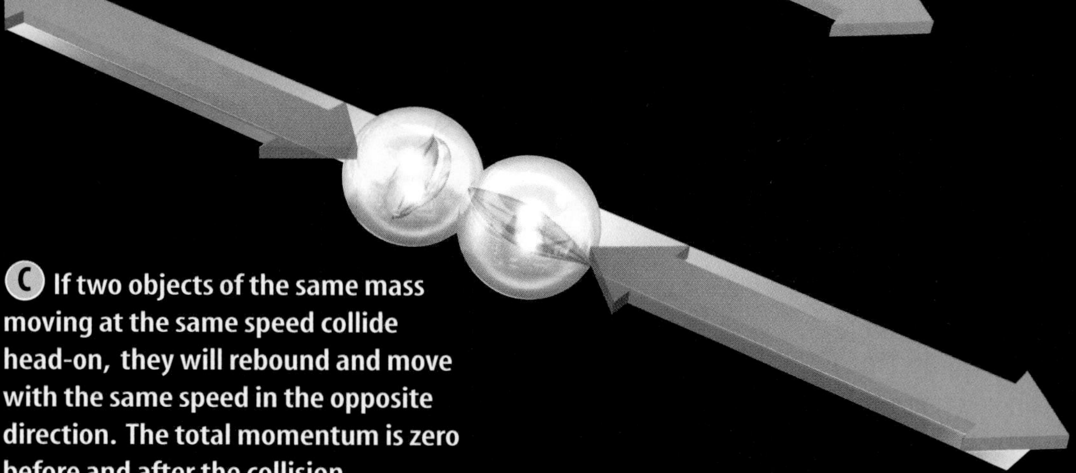

C If two objects of the same mass moving at the same speed collide head-on, they will rebound and move with the same speed in the opposite direction. The total momentum is zero before and after the collision.

Figure 17
When bumper cars collide, they bounce off each other, and momentum is transferred.

Bouncing Off In some types of collisions the objects involved, like the bumper cars in **Figure 17,** bounce off each other when they collide. The law of conservation of momentum can be used to help determine how these objects move after they collide. The results of collisions between two objects of various masses are shown in **Figure 16.**

For example, what happens if two objects of the same mass moving with the same speed collide head on? The objects reverse their direction of motion after the collision but still move with the same speed. What happens when one object directly hits an object at rest and both have the same mass? The first object transfers all of its momentum to the object at rest and comes to a stop.

Section 3 Assessment

1. When a player uses a golf club to hit a ball, how is momentum transferred?

2. What is the momentum of a 0.1-kg mass moving at 5 m/s?

3. A system of two balls has a momentum of 1 kg·m/s. Ball A has a momentum of −3 kg·m/s. What is the momentum of ball B?

4. **Think Critically** You watch a film in which one billiard ball rolls forward and hits another. After the collision the second billiard ball rolls away and the first one is motionless. Can you tell whether the film is being shown forward or in reverse?

Skill Builder Activities

5. **Predicting** Two balls of the same mass move toward each other with equal speeds and in the opposite direction. Predict how the balls will move after the collision if they collide and then stick together. Explain your answer. **For more help, refer to the** Science Skill Handbook.

6. **Solving One-Step Equations** A 0.2-kg ball is moving left at 3 m/s. It strikes a 0.5-kg ball that is at rest. Immediately after the collision, the 0.2-kg ball comes to a stop. How fast is the 0.5-kg ball moving if the momentum is conserved? **For more help, refer to the** Math Skill Handbook.

Activity

Collisions

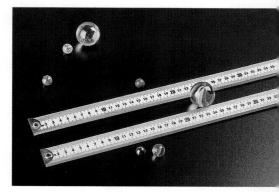

A collision occurs when a baseball bat hits a baseball or a tennis racket hits a tennis ball. What would happen if you hit a baseball with a table-tennis paddle, or a table-tennis ball with a baseball bat? How do the masses of colliding objects change the results of collisions?

What You'll Investigate
How does changing the size and number of marbles in a collision affect the collision?

Materials
small marbles (5)   metersticks (2)
large marbles (2)   tape

Goals
- **Compare and contrast** different collisions.
- **Determine** how the speeds after a collision depend on the masses of the colliding objects.

Safety Precautions 👓

Procedure

1. Tape the metersticks next to each other, slightly farther apart than the width of the large marbles. This limits the motion of the marbles to nearly a straight line.

2. Place a small marble in the center of the track formed by the metersticks. Place another small marble at one end of the track. Shoot this marble toward the small target marble by flicking it with your finger. Describe the collision.

3. Repeat step 2, replacing the two small marbles with the two large marbles.

4. Repeat step 2, replacing the small shooter marble with a large marble.

5. Repeat step 2, replacing the small target marble with a large marble.

6. Repeat step 2, replacing the small target marble with four small marbles that are touching.

7. Place two small marbles at opposite ends of the metersticks. Shoot the marbles toward each other and describe the collision.

8. Place two large marbles at opposite ends of the metersticks. Shoot the marbles toward each other and describe the collision.

9. Place a small marble and a large marble at opposite ends of the metersticks. Shoot the marbles toward each other and describe the collision.

Conclude and Apply

1. **Compare and contrast** the results of the various types of collisions.

2. In which collisions did the shooter marble change direction? How did the mass of the target marble compare with the shooter marble in these collisions?

Communicating Your Data

Make a chart showing your results. You might want to make before-and-after sketches, with short arrows to show slow movement and long arrows to show fast movement. **For more help, refer to the** Science Skill Handbook.

Activity
Design Your Own Experiment

Car Safety Testing

Imagine that you are a car designer. How can you create an attractive, fast car that is safe? When a car crashes, the passengers have inertia that can keep them moving. How can you protect the passengers from stops caused by sudden, head-on impacts?

Recognize the Problem

How can you design a car to win a race and protect the passenger in a head-on crash at the end of the race?

Form a Hypothesis

Develop a hypothesis about how to design a car to deliver a plastic egg quickly and safely through a race course and a crash at the end.

Goals
- **Construct** a fast car.
- **Design** a safe car that will protect a plastic egg from the effects of inertia when the car crashes.

Safety Precautions

Protect your eyes from possible flying objects.

Possible Materials
insulated foam meat trays or fast food trays
insulated foam cups
straws, narrow and wide
straight pins
tape
plastic eggs

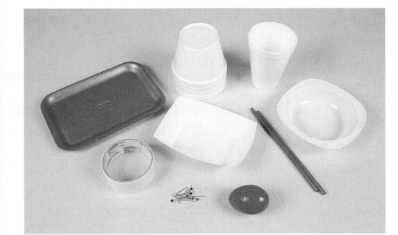

Test Your Hypothesis

Plan

1. Be sure your group has agreed on the hypothesis statement.

2. **Sketch** the design for your car. List the materials you will need. Remember that to make the car move smoothly, narrow straws will have to fit into the wider straws.

3. As a group, make a detailed list of the steps you will take to test your hypothesis.

4. Gather the materials you will need to carry out your experiment.

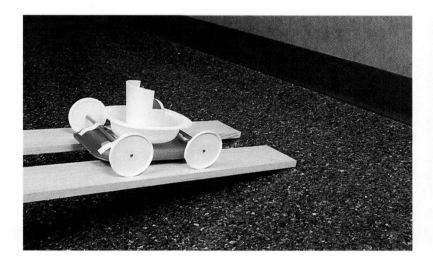

Do

1. Make sure your teacher approves your plan before you start. Include any changes suggested by your teacher in your plans.

2. Carry out the experiment as planned.

3. **Record** any observations that you made while doing your experiment. Include suggestions for improving your design.

Analyze Your Data

1. **Compare** your car design to the designs of the other groups. What made the fastest car fast? What slowed the slowest car?

2. **Compare** your car's safety features to those of the other cars. What protected the eggs the best? How could you improve the unsuccessful designs?

3. What effect would decreasing the speed of your car have on the safety of the egg?

Draw Conclusions

1. How did the best designs protect the egg?

2. If you were designing cars, what could you do to better protect passengers from sudden stops?

ommunicating
Your Data

Write a descriptive paragraph on how a car could be designed to protect its passengers effectively. Include a sketch of your designs.

What Goes Around Comes Around

The Story of Boomerangs

Picture this—A group of kids is gathered on the flat, yellow plain near their encampment in the Australian outback. One youth steps forward. With the flick of an arm, the youth flings a long, flat, angled stick that soars and spins into the sky. The stick's path curves until it returns—spinning right back into the thrower's hand. Another thrower steps forward with another stick. The contest goes on all afternoon.

That scene could be from today. Or, it could have taken place 10,000—or more—years ago. The kids were throwing boomerangs—elegantly curved sticks.

Because of how boomerangs are shaped and thrown, they always return to the thrower.

Archaeologists in Australia have unearthed boomerangs from 15,000 years ago. The boomerang developed from simple clubs that early people threw to stun and kill prey animals. These people became very good throwers and probably soon discovered that clubs with different shapes had different properties in the air. They gradually refined their designs into a throwing stick resembling today's boomerangs. As boomerangs became more refined, they also might have been used for fun.

Modern boomerangs are made from many things—from paper to wood.

Making a Comeback

There are many different types of boomerangs, but they have a few things in common, each of which is related to how a boomerang works. First, boomerangs are flat on the bottom and rounded on the top. This shape helps provide lift, just like it does for an airplane wing. When you hold a boomerang to throw it, the flat side is against your palm and your fingers curl around the curved side.

Second, boomerangs are angled. This makes the boomerang spin in the air. When it is thrown correctly, a boomerang spins at right angles to the ground, not horizontally like helicopter blades.

Today, using boomerangs for fun is a popular sport. World-class boomerangers compete at the World Boomerang Championships. Throwers compete in events such as Time Aloft, in which they try to keep the boomerang up in the air as long as possible. In Team Terror, a grueling relay race, throwers perform difficult throws and trick catches, such as behind-the-back or under-the-leg. They use ultra-modern boomerangs—some of which look nothing like the old-style classic boomerang. But they carry on an ancient Australian tradition of competing for the best throw.

CONNECTIONS Design Boomerangs are made from materials ranging from a piece of paper to expensive hardwood. Research at the library to find instructions for making boomerangs. After you and your friends build some boomerangs, have a competition of your own.

SCIENCE *Online*
For more information, visit
science.glencoe.com.

Reviewing Main Ideas

Section 1 What is motion?

1. An object is in motion if it is changing position.

2. Distance measures the length of the path that an object follows during its motion. Displacement is the change in position between the starting point and the ending point, as well as the direction from the start to end points.

3. Speed is a measure of how quickly the position of an object changes. Velocity includes the speed and direction of motion.

4. Speed can be calculated using the following equation:

$$\text{speed} = \frac{\text{distance}}{\text{time}}$$

5. A distance-time graph can be used to show motion. *Which object is moving the fastest?*

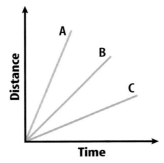

Section 2 Acceleration

1. Acceleration is a measure of how quickly velocity changes. It includes a direction.

2. An object is accelerating when it speeds up, slows down, or turns.

3. When an object speeds up or slows down, its acceleration can be calculated by

$$a = \frac{(s_f - s_i)}{t}$$

4. An object's acceleration can be determined from the speed-time graph. *During what time intervals is the object speeding up? Slowing down? How can you tell?*

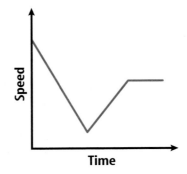

Section 3 Momentum

1. The momentum of an object is a measure of how hard it is to stop the object. Momentum is the product of mass and velocity. It has a direction.

2. Momentum is transferred from one object to another in a collision.

3. According to the law of conservation of momentum, the total amount of momentum of a group of objects does not change unless outside forces act on the objects. *How could you determine the total momentum of these balls? Would it change after they collide? Why or why not?*

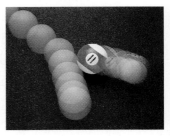

After You Read

FOLDABLES
Reading & Study
Skills

Use each vocabulary word on your Foldable in a sentence about motion and write it next to the definition of the word.

Visualizing Main Ideas

Complete the following table comparing different descriptions of motion.

Describing Motion

Quantity	Definition	Direction?
Distance	length of path traveled	no
Displacement	direction and change in position	yes
Speed		no
Velocity	rate of change in position and direction	
Acceleration		
Momentum		yes

Vocabulary Review

Vocabulary Words

a. acceleration
b. average speed
c. inertia
d. instantaneous speed
e. law of conservation
   of momentum
f. mass
g. momentum
h. speed
i. velocity

Using Vocabulary

Explain the relationship between each pair of words.

1. speed, velocity

2. velocity, acceleration

3. velocity, momentum

4. momentum, law of conservation of momentum

5. mass, momentum

6. mass, inertia

7. momentum, inertia

8. average speed, instantaneous speed

THE PRINCETON REVIEW — Study Tip

Pay attention to the chapter's illustrations. Try to figure out exactly the main point each picture is trying to stress.

Checking Concepts

Choose the word or phrase that best answers the question.

1. What measures the quantity of matter?
   A) speed
   B) weight
   C) acceleration
   D) mass

2. A 2-kg ball has a momentum of 10 kg·m/s. What is its speed?
   A) 1 m/s
   B) 5 m/s
   C) 10 m/s
   D) 20 m/s

3. Which of the following is NOT an example of acceleration?
   A) a leaf falling at constant speed
   B) a car slowing down
   C) a skater spinning
   D) a dog running faster and faster

4. Speed is related to change in which of the following?
   A) distance
   B) momentum
   C) velocity
   D) acceleration

5. A parked car is hit by a moving car, and the two cars stick together. How does the speed of the combined cars compare to the speed of the car before the collision?
   A) Combined speed is the same.
   B) Combined speed is greater.
   C) Combined speed is smaller.
   D) Any of these could be true.

6. A car travels for half an hour at 40 km/h. How far does it travel?
   A) 10 km
   B) 20 km
   C) 40 km
   D) 80 km

7. What is a measure of inertia?
   A) weight
   B) gravity
   C) momentum
   D) mass

8. What is 18 cm/h north an example of?
   A) speed
   B) velocity
   C) acceleration
   D) momentum

9. Ball A bumps into ball B. Which of the following is the same before and after the collision?
   A) the momentum of ball A
   B) the momentum of ball B
   C) the sum of the momentums
   D) the difference in the momentums

10. What is velocity change divided by time change?
    A) speed
    B) displacement
    C) momentum
    D) acceleration

Thinking Critically

11. You run 100 m in 25 s. If you later run the same distance in less time, does your average speed increase or decrease? Explain.

12. A car drives from point A to point B. Can its displacement be greater than the total distance that it traveled? Explain.

13. The Moon moves around Earth at close to constant speed. Does this mean it is not accelerating? Explain.

14. When a wrecking ball hits a wall, it is not moving fast. How can it have enough momentum to knock down a wall?

Developing Skills

15. **Predicting** A rocket accelerates at 12 m/s^2 for 8 s. If it starts at rest, what is its final speed?

16. **Measuring in SI** Measure the width of your desk. Time a pen rolling across the desk, and find the speed of the pen.

17. Making Models The molecules in a gas are modeled as colliding balls. If the molecules all have the same mass, explain what can happen when a fast-moving molecule hits a slow-moving molecule. Include a sketch.

18. Recognizing Cause and Effect Roll one marble into the end of a line of stationary marbles. What do you observe? Why do you think this happened?

19. Making and Using Graphs Use the speed-time graph below. What is the acceleration between $t = 0$ and $t = 3$?

Speed versus Time

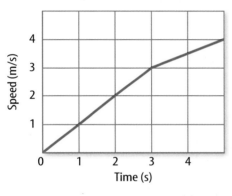

Time (s)

Performance Assessment

20. Demonstration Set up a racetrack and make rules for the type of motion allowed. Demonstrate how to measure distance, measure time, and calculate speed accurately.

21. Poster Make a poster showing how motion graphs represent acceleration and constant speed. Use both speed-time and distance-time graphs.

TECHNOLOGY

Go to the Glencoe Science Web site at **science.glencoe.com** or use the **Glencoe Science CD-ROM** for additional chapter assessment.

THE PRINCETON REVIEW **Test Practice**

Two students are comparing the performance of different electric toy cars by measuring the top speed each car can reach. They also calculate the momentum of each car at its top speed. The table below summarizes their results.

Performance of Toy Cars

Car	Mass (kg)	Velocity (m/s)	Momentum (kg·m/s)
A	1.0	0.2	0.2
B	1.0	0.5	0.5
C	0.5	1.6	0.8
D	3.0	0.2	0.6
E	2.0	0.4	0.8

Study the table and answer the following questions.

1. According to the table, which statement best describes how the students calculated the momentum of the cars?
  A) momentum = mass + velocity
  B) momentum = mass − velocity
  C) momentum = mass × velocity
  D) momentum = mass/velocity

2. About how much greater is the momentum of the car with the least mass than that of the car with the greatest mass?
  F) 0.2 kg·m/s
  G) 0.4 kg·m/s
  H) 1.0 kg·m/s
  J) 0.5 kg·m/s

Force and Newton's Laws

This train can move! Japan's experimental train, the MLX01, can travel at speeds of more than 500 km/h. It can move this fast because the engineers who built it understood the laws of motion that Isaac Newton first proposed more than 300 years ago. In this chapter, you'll learn about force and Newton's laws of motion. You'll learn why some objects move and why some stay still, and how objects exert forces on each other.

What do you think?

Science Journal Look at the picture below with a classmate. Discuss what this might be or what is happening. Here's a hint: *It's a delicate balancing act.* Write your answer or best guess in your Science Journal.

EXPLORE ACTIVITY

Imagine being on a bobsled team speeding down an icy run. The force of gravity causes the sled to accelerate as it speeds down the course in a blur. You and your team use your bodies, brakes, and the steering mechanism to exert forces to change the sled's motion, causing it to slow down or turn. The motion of the sled as it speeds up, slows down, and turns can be explained with Newton's laws of motion. These laws tell how forces cause the motion of an object to change.

Analyze motion on a ramp

1. Lean two metersticks on three books as shown to the right. This is your ramp.

2. Tap a marble so it rolls up the ramp. Measure how far up the ramp it travels before rolling back.

3. Repeat step 2 using two books, one book, and zero books. The same person should tap with the same force each time.

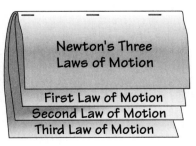

Observe

Make a table to record the motion of the marble for each ramp height. What would happen if the ramp were perfectly smooth and level?

Before You Read

FOLDABLES
Reading & Study Skills

Making an Organizational Study Fold When information is grouped into clear categories, it is easier to make sense of what you are learning. Make the following Foldable to help you organize your thoughts about Newton's Three Laws of Motion.

1. Stack two sheets of paper in front of you so the short side of both sheets is at the top.

2. Slide the top sheet up so that about four centimeters of the bottom sheet show.

3. Fold both sheets top to bottom to form four tabs and staple along the topfold as shown.

4. Label each flap *Newton's Three Laws of Motion, First Law of Motion, Second Law of Motion,* and *Third Law of Motion* as shown.

5. As you read the chapter, record what you learn about the laws of motion under the tabs.

> Newton's Three
> Laws of Motion
>
> First Law of Motion
> Second Law of Motion
> Third Law of Motion

① Newton's First Law

What You'll Learn

- **Identify** forces at work.
- **Distinguish** between balanced and net forces.
- **Demonstrate** Newton's first law of motion.
- **Explain** how friction works.

Vocabulary
force
net force
balanced forces
unbalanced forces
Newton's first law of motion
friction

Why It's Important
Newton's first law helps you understand why objects slow down and stop.

Force

A soccer ball sits on the ground, motionless, until you kick it. Your science book sits on the table until you pick it up. If you hold your book above the ground, then let it go, gravity pulls it to the floor. In every one of these cases, the motion of the ball or book was changed by something pushing or pulling on it. An object will speed up, slow down, or turn only if something is pushing or pulling on it.

A **force** is a push or a pull. Examples of forces are shown in **Figure 1.** Think about throwing a ball. Your hand exerts a force on the ball, and the ball accelerates forward until it leaves your hand. After the ball leaves your hand, gravity's force on it causes its path to curve downward. When the ball hits the ground, the ground exerts a force, stopping the ball.

A force can be exerted in different ways. For instance, a paper clip can be moved by the force a magnet exerts, the pull of Earth's gravity, or the force you exert when you pick it up. These are all examples of forces acting on the paper clip.

Figure 1
A force is a push or a pull.

B The magnet on the crane pulls the pieces of scrap metal upward.

A This golf club exerts a force by pushing on the golf ball.

A This door is not moving because the forces exerted on it are equal and in opposite directions.

B The door is closing because the force pushing the door closed is greater than the force pushing it open.

Combining Forces More than one force can act on an object at the same time. If you hold a paper clip near a magnet, you, the magnet, and gravity all exert forces on the paper clip. The combination of all the forces acting on an object is the **net force.** When more than one force is acting on an object, the net force determines the motion of the object. In this example, the paper clip is not moving, so the net force is zero.

How do forces combine to form the net force? If the forces are in the same direction, they add together to form the net force. If two forces are in opposite directions, then the net force is the difference between the two forces, and it is in the direction of the larger force.

Balanced and Unbalanced Forces A force can act on an object without causing it to accelerate if other forces cancel the push or pull of the force. Look at **Figure 2.** If you and your friend push on a door with the same force in opposite directions, the door does not move. Because you both exert forces of the same size in opposite directions on the door, the two forces cancel each other. Two or more forces exerted on an object are **balanced forces** if their effects cancel each other and they do not cause a change in the object's motion. If the forces on an object are balanced, the net force is zero. If the forces are **unbalanced forces,** their effects don't cancel each other. Any time the forces acting on an object are unbalanced, the net force is not zero and the motion of the object changes.

Figure 2
When the forces on an object are balanced, no change in motion occurs, but when the forces on an object are unbalanced, a change in motion does occur.

Life Science
INTEGRATION

Whether you run, jump, or sit, forces are being exerted on different parts of your body. Biomechanics is the study of how the body exerts forces and how it is affected by forces acting on it. Research how biomechanics has been used to reduce job-related injuries. Write a paragraph on what you've learned in your Science Journal.

Newton's First Law of Motion

If you stand on a skateboard and someone gives you a push, then you and your skateboard will start moving. You began to move when the force was applied. An object at rest—like you on your skateboard—remains at rest unless an unbalanced force acts on it and causes it to move.

Because a force had to be applied to make you move when you and your skateboard were at rest, you might think that a force has to be applied continually to keep an object moving. Surprisingly, this is not the case. An object can be moving even if the net force acting on it is zero.

Newton's first law of motion describes how an object moves when no net force is acting on it. According to **Newton's first law of motion,** if there is no net force acting on an object the object remains at rest, or if the object is already moving, it continues to move in a straight line with constant speed.

The Italian scientist Galileo Galilei, who lived from 1564 to 1642, was one of the first to understand that a force doesn't need to be constantly applied to an object to keep it moving. Galileo's ideas helped Isaac Newton to better understand the nature of motion. Newton was able to explain the motion of objects in three rules called Newton's laws of motion.

Friction

Galileo realized the motion of an object doesn't change until an unbalanced force acts on it. Every day you see moving objects come to a stop. The force that brings nearly everything to a stop is **friction,** which is the force that acts to resist sliding between two touching surfaces, as shown in **Figure 3.**

Friction is why you never see objects moving with constant velocity unless a net force is applied. Friction is the force that eventually brings your skateboard to a stop unless you keep pushing on it. Friction always acts on objects that are sliding or moving through air or water.

Figure 3
When two objects in contact try to slide past each other, friction keeps them from moving or slows them down.

A Without friction, the rock climber would slide down the rock.

Force due to friction

Force due to friction

Force due to friction

Force due to gravity

B Friction slows down this sliding baseball player.

Force due to friction

Opposing Sliding Although several different forms of friction exist, they all have one thing in common. If two objects are in contact, frictional forces always try to prevent one object from sliding on the other object. If you rub your hand against a tabletop, you can feel the friction push against the motion of your hand. If you rub the other way, you can feel the direction of friction change so it is again acting against your hand's motion. Friction always will slow an object down.

✔ **Reading Check** *What do the different forms of friction have in common?*

Older Ideas About Motion It took a long time for people to understand motion. One reason was that people did not understand the behavior of friction or understand that friction was a force. Because friction causes moving objects to stop, people thought the natural state of an object was to be at rest. For an object to be in motion, something had to be pushing or pulling it continuously. As soon as the force stopped, nature would bring the object to rest.

Galileo understood that an object in constant motion is as natural as an object at rest. It was usually friction that made moving objects slow down and eventually come to a stop. To keep an object moving, a force had to be applied to overcome the effects of friction. If friction could be removed, an object in motion would continue to move in a straight line with constant speed. **Figure 4** shows motion where there is almost no friction.

SCIENCE
Online

Research Visit the Glencoe Science Web site at **science.glencoe.com** for more information about the lives of Galileo and Newton. Communicate to your class what you learn.

Figure 4
In an air hockey game, the puck floats on a layer of air, so that friction is almost eliminated. As a result, the puck moves in a straight line with nearly constant speed after it's been hit. *How would the puck move if there was no layer of air?*

Observing Friction

Procedure

1. Lay a **bar of soap,** a **flat eraser,** and a **key** side by side on one end of a **hard-sided notebook.**
2. At a constant rate, slowly lift the end of notebook with objects on it. Note the order in which the objects start sliding.

Analysis

1. For which object was static friction the greatest? For which object was it the smallest? Explain, based on your observations.
2. Which object slid the fastest? Which slid the slowest? Explain why there is a difference in speed.
3. How could you increase and decrease the amount of friction between two materials?

Static Friction If you've ever tried pushing something heavy, like a refrigerator, you might have discovered that nothing happened at first. Then as you push harder and harder, the object suddenly will start to move. When you first start to push, friction between the heavy refrigerator and the floor opposes the force you are exerting and the net force is zero. The type of friction that prevents an object from moving when a force is applied is called static friction.

Static friction is caused by the attraction between the atoms on the two surfaces that are in contact. This causes the surfaces to stick or weld together where they are in contact. Usually, as the surface gets rougher and the object gets heavier, the force of static friction will be larger. To move the object, you have to exert a force large enough to break the bonds holding two surfaces together.

Sliding Friction While static friction keeps an object at rest, sliding friction slows down an object that slides. If you push an object across a room, you notice the sliding friction between the bottom of the object and the floor. You have to keep pushing to overcome the force of sliding friction. Sliding friction is due to the microscopic roughness of two surfaces, as shown in **Figure 5.** A force must be applied to move the rough areas of one surface past the rough areas of the other. The brake pads in a car use sliding friction against the wheels to slow the car. Bicycle brakes, shown in **Figure 6A,** work the same way.

✓ **Reading Check** *What is the difference between static friction and sliding friction?*

Figure 5
Microscopic roughness, even on surfaces that seem smooth, such as the tray and metal shelf, causes sliding friction. *What do you think a lubricant does?*

Figure 6
A bicycle uses sliding friction and rolling friction.

A Sliding friction is used to stop this bicycle tire. Friction between the brake pads and the wheel brings the wheel to a stop.

Wheel turning

Force due to friction

B Rolling friction with the ground pushes the bottom of the bicycle tire, so it rolls forward.

Force due to friction

Rolling Friction Another type of friction, rolling friction, is needed to make a wheel or tire turn. Rolling friction occurs between the ground and the part of the tire touching the ground, as shown in **Figure 6B.** Rolling friction keeps the tire from slipping on the ground. If the bicycle tires are rolling forward, rolling friction exerts the force on the tires that pushes the bicycle forward.

It's usually easier to pull a load on a wagon or cart that has wheels rather than to drag the load along the ground. This is because rolling friction between the wheels and the ground is less than the sliding fiction between the load and the ground.

Section 1 Assessment

1. A car turns to the left at 20 km/h. Is a force acting on the car? Explain.

2. Explain why friction made it difficult to discover Newton's first law of motion.

3. If the net force on an object is zero, are the forces acting on it balanced or unbalanced? Explain.

4. What makes static friction increase?

5. **Think Critically** In the following situations, are the forces balanced or unbalanced? How can you tell?
   a. You push a box until it moves.
   b. You push a box at a constant rate.
   c. You stop pushing a box, and it stops.

Skill Builder Activities

6. **Comparing and Contrasting** Compare and contrast static friction, sliding friction, and rolling friction. **For more help, refer to the** Science Skill Handbook.

7. **Communicating** Most of the meteors that reach Earth's atmosphere burn up on the way down. Friction between the meteor and the atmosphere produces a great deal of heat. Research how the space shuttle is protected from friction when it reenters Earth's atmosphere. Report your findings in your Science Journal. **For more help, refer to the** Science Skill Handbook.

Newton's Second Law

As You Read

What You'll Learn

■ **Explain** Newton's second law of motion.

■ **Explain** why the direction of force is important.

Vocabulary

Newton's second law of motion
weight

Why It's Important

Newton's second law of motion explains how any object, from a swimmer to a satellite, moves when acted on by forces.

Force and Acceleration

When you go shopping in a grocery store and push a cart, you exert a force to make the cart move. If you want to slow down or change the direction of the cart, a force is required to do this, as well. Would it be easier for you to stop a full or empty grocery cart suddenly, as in **Figure 7?** When the motion of an object changes, the object is accelerating. Acceleration occurs any time an object speeds up, slows down, or changes its direction of motion. Newton's second law describes how forces cause an object's motion to change.

Newton's second law of motion connects force, acceleration, and mass. According to the second law of motion, an object acted upon by a force will accelerate in the direction of the force. The acceleration is given by the following equation

$$\text{acceleration} = \frac{\text{net force}}{\text{mass}}$$

$$a = \frac{F_{\text{net}}}{m}$$

In this equation, a is the acceleration, m is the mass, and F_{net} is the net force. If both sides of the above equation are multiplied by the mass, the equation can be written this way:

$$F_{\text{net}} = ma$$

✓ **Reading Check** *What is Newton's second law?*

Figure 7
The force needed to change the motion of an object depends on its mass. *Which grocery cart would be easier to stop suddenly?*

Units of Force Force is measured in newtons, abbreviated N. Because the SI unit for mass is the kilogram (kg) and acceleration has units of meters per second squared (m/s^2), 1 N also is equal to 1 kg·m/s^2. In other words, to calculate a force in newtons from the equation shown on the prior page, the mass must be given in kg and the acceleration in m/s^2.

Gravity

One force that you are familiar with is gravity. Whether you're coasting down a hill on a bike or a skateboard or jumping into a pool, gravity is at work pulling you downward. Gravity also is the force that causes Earth to orbit the Sun and the Moon to orbit Earth.

What is gravity? The force of gravity exists between any two objects that have mass. Gravity always is attractive and pulls objects toward each other. A gravitational attraction exists between you and every object in the universe that has mass. However, the force of gravity depends on the mass of the objects and the distance between them. The gravitational force becomes weaker the farther apart the objects are and also decreases as the masses of the objects involved decrease.

For example, there is a gravitational force between you and the Sun and you and Earth. The Sun is much more massive than Earth, but is so far away that the gravitational force between you and the Sun is too weak to notice. Only Earth has enough mass and is close enough to exert a noticeable gravitational force on you. The force of gravity between you and Earth is about 1,650 times greater than between you and the Sun.

Astronomy
INTEGRATION

A black hole is a star that has collapsed so that all its mass is compressed into a small region that may be less than 10 km in diameter. Near a black hole, the force of gravity is much stronger than the force of gravity near Earth. Research some of the unusual phenomena that occur near black holes.

Weight The force of gravity causes all objects near Earth's surface to fall with an acceleration of 9.8 m/s^2. By Newton's second law, the gravitational force on any object near Earth's surface is:

$$F = ma = m \times (9.8 \text{ m/s}^2)$$

This gravitational force also is called the weight of the object. Your **weight** on Earth is the gravitational force between you and Earth. Your weight would change if you were standing on a different planet than Earth, as shown in **Table 1**. Your weight on a different planet is the gravitational force between you and the planet.

Table 1 Weight of 60 kg Person on Various Planets

Place	Weight in Newtons If Your Mass Were 60 kg	Percent of Your Weight on Earth
Mars	221	38
Earth	587	100
Jupiter	1,387	236
Pluto	39	0.7

Figure 8
The girl is speeding up because she is being pushed in the same direction that she is moving.

Applied force

Direction of motion

Figure 9
The boy is slowing down because the force exerted by his feet is in the opposite direction of his motion.

Direction of motion

Force due to friction

Weight and Mass Weight and mass are different. Weight is a force, just like the push of your hand is a force, and is measured in newtons. When you stand on a bathroom scale, you are measuring the pull of Earth's gravity—a force. However, mass is the amount of matter in an object, and doesn't depend on location. A book with a mass of 1 kg has a mass of 1 kg on Earth or on Mars. However, the weight of the book would be different on Earth and Mars. The two planets would exert a different gravitational force on the book.

Using Newton's Second Law

How does Newton's second law determine how an object moves when acted upon by forces? The second law tells how to calculate the acceleration of an object if its mass and the forces acting on it are known. You may remember that the motion of an object can be described by its velocity. The velocity tells how fast an object is moving and in what direction. Acceleration tells how velocity changes. If the acceleration of an object is known, then the change in velocity can be determined.

Speeding Up Think about a soccer ball sitting on the ground. If you kick the ball, it starts moving. You exert a force on the ball, and the ball accelerates only while your foot is in contact with the ball. If you look back at all of the examples of objects speeding up, you'll notice that something is pushing or pulling the object in the direction it is moving, as in **Figure 8.** The direction of the push or pull is the direction of the force. It also is the direction of the acceleration.

Calculating Acceleration Newton's second law of motion can be used to calculate acceleration. For example, suppose you pull a 10-kg sled so that the net force on the sled is 5 N. The acceleration can be found as follows:

$$a = \frac{F_{net}}{m} = \frac{5 \text{ N}}{10 \text{ kg}} = 0.5 \text{ m/s}^2$$

The sled keeps accelerating as long as you keep pulling on it. The acceleration does not depend on how fast the sled is moving. It depends only on the net force and the mass of the sled.

Slowing Down If you wanted to slow down an object, you would have to push or pull it against the direction it is moving. An example is given in **Figure 9.** Here the force is opposite to the velocity or the direction of motion.

Suppose you push a book so it slides across a tabletop. You exert a force on the book when your hand is in contact with it, and the book speeds up. Sliding friction also acts on the book as it starts to move. After the book is no longer in contact with your hand, friction acts in the opposite direction to the book's motion. This causes the book to slow down and come to a stop.

Field GUIDE

How does acceleration affect how you feel on a roller coaster? To find out more about acceleration and amusement park rides, see the **Amusement Park Rides Field Guide** at the back of the book.

Math Skills Activity

Calculating Force Using Newton's Second Law

Example Problem
    A car with a mass of 1,500 kg has an acceleration of 3 m/s^2. Find the force acting on the car.

Solution

1 *This is what you know:*   acceleration: $a = 3$ m/s^2
                                          mass: $m = 1,500$ kg

2 *This what you need to find:*   Force: F

3 *This is the equation you need to use:* $F = ma$

4 *Substitute the known values:* $F = (1,500 \text{ kg}) \times (3 \text{ m/s}^2) = 4,500$ N

    Check your answer by dividing the force you calculate by the acceleration that was given. Do you calculate the same mass that was given?

Practice Problem

    You throw a baseball with a mass of 0.15 kg so it has an acceleration of 40 m/s^2. Find the force you exerted on the baseball.

For more help, refer to the Math Skill Handbook.

Figure 10
When the ball is thrown, it doesn't keep moving in a straight line. Gravity exerts a force downward that makes it move in a curved path.

Direction of motion

Force due to gravity

Turning Sometimes forces and motion are not in a straight line. If a net force acts at an angle to the direction an object is moving, the object will follow a curved path. The object might be going slower, faster, or at the same speed after it turns.

For example, when you shoot a basketball, the ball doesn't continue to move in a straight line after it leaves your hand. Instead it starts to curve downward, as shown in **Figure 10.** The force of gravity pulls the ball downward. The ball's motion is a combination of its original motion and the downward motion due to gravity. This causes the ball to move in a curved path.

Circular Motion

A rider on a merry-go-round ride moves in a circle. This type of motion is called circular motion. If you are in circular motion, your direction of motion is constantly changing. This means you are constantly accelerating. According to Newton's second law of motion, if you are constantly accelerating, there must be a force acting on you the entire time.

Think about an object on the end of a string whirling in a circle. The force that keeps the object moving in a circle is exerted by the string. The string pulls on the object to keep it moving in a circle. The force exerted by the string is the centripetal force and always points toward the center of the circle. In circular motion the centripetal force is always perpendicular to the motion.

Satellite Motion Objects that orbit Earth are satellites of Earth. Satellites go around Earth in nearly circular orbits, with the centripetal force being gravity. If gravity is pulling satellites toward Earth, why doesn't a satellite fall to Earth like a baseball does? Actually, a satellite is falling to Earth just like a baseball.

Suppose Earth were perfectly smooth with no mountains or hills. Imagine you throw a baseball horizontally. Gravity pulls the baseball downward so it travels in a curved path. If the baseball is thrown faster, its path is less curved, and it travels farther before it hits the ground, as shown in **Figure 11.** If the baseball were traveling fast enough, as it fell, its curved path would follow the curve of Earth's surface. Then the baseball would never hit the ground. Instead, it would continue to fall around Earth.

Satellites in orbit are being pulled toward Earth just as baseballs are. The difference is that satellites are moving so fast horizontally that Earth's surface curves downward at the same rate that the satellites are falling downward. The speed at which a object must move to go into orbit near Earth's surface is about 8 km/s, or about 29,000 km/h.

To place a satellite into orbit, a rocket carries the satellite to the desired height. Then the rocket fires again to give the satellite the horizontal speed it needs to stay in orbit.

Figure 11
The faster a ball is thrown, the farther it travels before gravity pulls it to Earth. If the ball is traveling fast enough, Earth's surface curves away from it as fast as it falls downward. Then the ball never hits the ground.

Air Resistance

Whether you are walking, running, or biking, air is pushing against you. This push is air resistance. Air resistance is a form of friction that acts to slow down any object moving in the air. Air resistance gets larger as an object moves faster.

When an object falls it speeds up as gravity pulls it downward. At the same time, the force of air resistance pushing up on the object is increasing as the object moves faster. Finally, the upward air resistance force becomes large enough to equal the downward force of gravity.

When the air resistance force equals the weight, the net force on the object is zero. By Newton's second law, the object's acceleration then is zero, and its speed no longer increases. The constant speed a falling object reaches when air resistance balances the force of gravity is the terminal velocity.

B

A

Figure 12
Sky divers can change their air resistance by changing the position of their arms and legs.
A In a spread-eagle position, the air resistance of the sky diver is greater. **B** With the legs closed and the arms tucked back against the body, the sky diver's shape is narrower and the air resistance is less.

Air Resistance and Shape

The amount of air resistance depends on the object's shape, as well as its speed. Moving at the same speed, the air resistance on a pointed, narrow object is less than on a broad, flat object, such as a leaf or a piece of paper. A falling sky diver in a spread-eagle position, as shown in **Figure 12A,** might reach a terminal velocity of about 200 km/h. But with the arms tucked backward and the legs closed, air resistance is less, and the skydiver might reach a terminal velocity of over 300 km/h. When the skydiver opens the parachute, the force of air resistance on an open parachute is so large that the skydiver's terminal velocity quickly is reduced to about 20 km/h.

Section 2 Assessment

1. A human cannonball with a mass of 80 kg is fired out of a cannon with a force of 2,400 N. Find the acceleration.

2. A bike rider traveling at 20 km/h on a flat roadway stops pedaling. Make a diagram showing the forces acting on the coasting bike and rider. Using Newton's second law, explain how the bike's motion will change.

3. Suppose you were in a spaceship traveling away from Earth. How would your weight change as you moved farther from Earth?

4. What happens when the air resistance force equals the weight of a falling object?

5. **Think Critically** Explain how you can determine the direction of a force by watching an object's change in motion.

Skill Builder Activities

6. **Drawing Conclusions** Three students are pushing on a box. Two students are pushing on the left side, and one is pushing on the right side. One student on the left pushes with a force of 10 N and the other pushes with a force of 15 N. The student of the right pushes with a force of 20 N. In what direction will the box move? Explain your answer. **For more help, refer to the** Science Skill Handbook.

7. **Solving One-Step Equations** A 1-kg ball is moving at 2 m/s. A force stops the ball in 4 s. Find the acceleration of the ball by dividing the change in speed on the ball by the time needed to stop. Then find the force. **For more help, refer to the** Math Skill Handbook.

Newton's Third Law

Action and Reaction

Newton's first two laws of motion explain how the motion of a single object changes. If the forces acting on the object are balanced, the object will remain at rest or stay in motion with constant velocity. If the forces are unbalanced, the object will accelerate in the direction of the net force. Newton's second law tells how to calculate the acceleration, or change in motion, of an object if the net force acting on it is known.

Newton's third law describes something else that happens when one object exerts a force on another object. Suppose you push on a wall. It may surprise you to learn that if you push on a wall, the wall also pushes on you. According to **Newton's third law of motion,** forces always act in equal but opposite pairs. Another way of saying this is for every action, there is an equal but opposite reaction. This means that when you push on a wall, the wall pushes back on you with a force equal in strength to the force you exerted. When one object exerts a force on another object, the second object exerts the same size force on the first object, as shown in **Figure 13.**

As You Read

What You'll Learn

■ **Identify** the relationship between the forces that objects exert on each other.

Vocabulary
Newton's third law of motion

Why It's Important
Newton's third law can help you predict how objects will affect one another.

Force on jack due to car

Force on car due to jack

Figure 13
The car jack is pushing up on the car with the same amount of force with which the car is pushing down on the jack.

Figure 14
In this collision, the first car exerts a force on the second. The second exerts the same force in the opposite direction on the first car. *Which car do you think accelerates more?*

Figure 15
When the child pushes against the wall, the wall pushes against the child.

Action and Reaction Forces Don't Cancel The forces exerted by two objects on each other are often called an action-reaction force pair. Either force can be considered the action force or the reaction force. You might think that because action-reaction forces are equal and opposite that they cancel. However, action and reaction force pairs don't cancel because they act on different objects. Forces can cancel only if they act on the same object.

For example, imagine you're driving a bumper car and are about to ram a friend in another car, as shown in **Figure 14.** When the two cars collide, your car pushes on the other car. By Newton's third law, that car pushes on your car with the same force, but in the opposite direction. This force causes you to slow down. One force of the action-reaction force pair is exerted on your friend's car, and the other force of the force pair is exerted on your car. Another example of an action-reaction pair is shown in **Figure 15.**

You constantly use action-reaction force pairs as you move about. When you jump, you push down on the ground. The ground then pushes up on you. It is this upward force that pushes you into the air. **Figure 16** shows some examples of how Newton's laws of motion are demonstrated in sporting events.

Reaction force
Action force

Life Science
INTEGRATION

Birds and other flying creatures also use Newton's third law. When a bird flies, its wings push in a downward and a backward direction. This pushes air downward and backward. By Newton's third law, the air pushes back on the bird in the opposite directions—upward and forward. This force keeps a bird in the air and propels it forward.

NATIONAL GEOGRAPHIC VISUALIZING NEWTON'S LAWS IN SPORTS

Figure 16

Although it is not obvious, Newton's laws of motion are demonstrated in sports activities all the time. According to the first law, if an object is in motion, it moves in a straight line with constant speed unless a net force acts on it. If an object is at rest, it stays at rest unless a net force acts on it. The second law states that a net force acting on an object causes the object to accelerate in the direction of the force. The third law can be understood this way—for every action force, there is an equal and opposite reaction force.

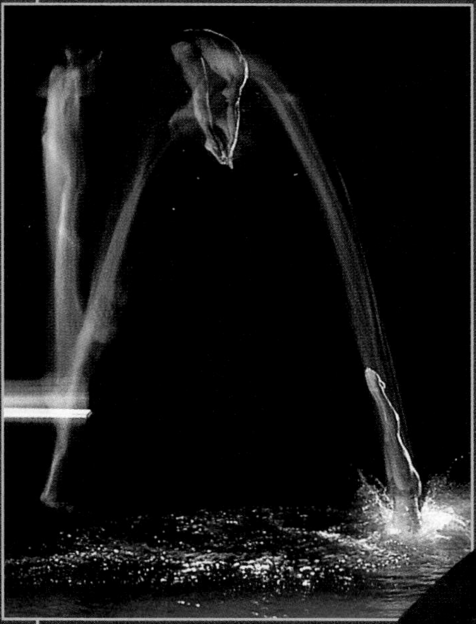

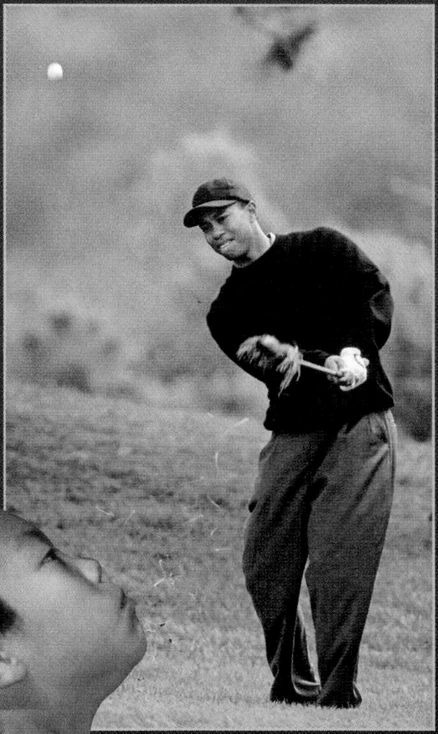

◀ **NEWTON'S SECOND LAW** As Tiger Woods hits a golf ball, he applies a force that will drive the ball in the direction of that force—an example of Newton's second law.

▲ **NEWTON'S FIRST LAW** According to Newton's first law, the diver does not move in a straight line with constant speed because of the force of gravity.

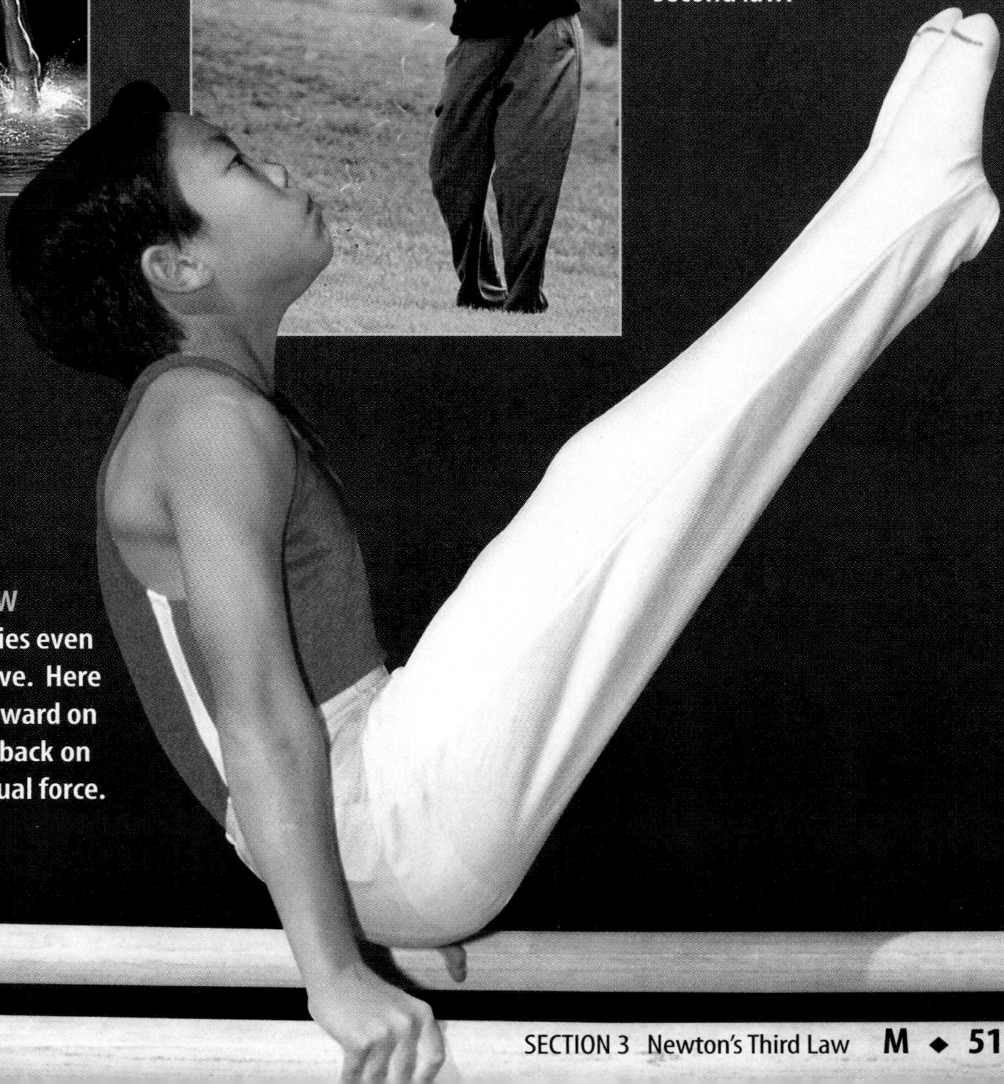

▶ **NEWTON'S THIRD LAW** Newton's third law applies even when objects do not move. Here a gymnast pushes downward on the bars. The bars push back on the gymnast with an equal force.

Figure 17

The force of the ground on your foot is equal and opposite to the force of your foot on the ground. If you push back harder, the ground pushes forward harder.

How do astronauts live under conditions of weightlessness? To find out more about the effects of weightlessness, see the **Living in Space Field Guide** at the end of the book.

Large and Small Objects Sometimes it's easy not to notice an action-reaction pair is because one of the objects is often much more massive and appears to remain motionless when a force acts on it. It has so much inertia, or tendency to remain at rest, that it hardly accelerates. Walking is a good example. When you walk forward, you push backward on the ground. Your shoe pushes Earth backward, and Earth pushes your shoe forward, as shown in **Figure 16.** Earth has so much mass compared to you that it does not move noticeably when you push it. If you step on something that has less mass than you do, like a skateboard, you can see it being pushed back.

A Rocket Launch The launching of a space shuttle is a spectacular example of Newton's third law. Three rocket engines supply the force, called thrust, that lifts the rocket. When the rocket fuel is ignited, a hot gas is produced. As the gas molecules collide with the inside engine walls, the walls exert a force that pushes them out of the bottom of the engine, as shown in **Figure 18.** This downward push is the action force. The reaction force is the upward push on the rocket engine by the gas molecules. This is the thrust that propels the rocket upward.

Gas particles

Engine compartment

Figure 18

Newton's third law enables a rocket to fly. The rocket pushes the gas molecules downward, and the gas molecules push the rocket upward.

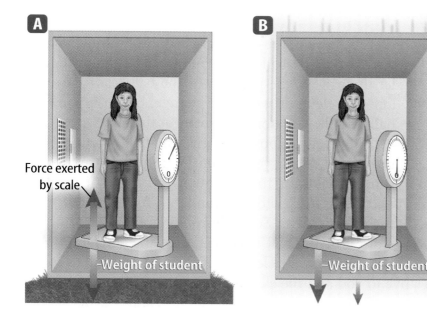

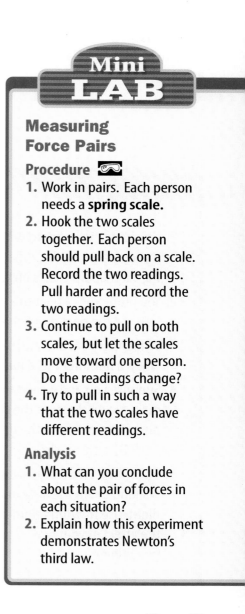

Figure 19
Your weight measured by a scale changes when you are falling. **A** When you stand on a scale on Earth, the reading on the scale is your weight. **B** If you were to stand on a scale in a falling elevator, the scale would read zero.

In figure:
A
Force exerted by scale
–Weight of student

B
–Weight of student

Weightlessness

You may have seen pictures of astronauts floating inside a space shuttle as it orbits Earth. The astronauts are said to be weightless, as if Earth's gravity were no longer pulling on them. Yet the force of gravity on the shuttle is still about 90 percent as large as at Earth's surface. Newton's laws of motion can explain why the astronauts float as if there were no forces acting on them.

Measuring Weight Think about how you measure your weight. When you stand on a scale, your weight pushes down on the scale and causes the springs in the scale to compress. The scale pointer moves from zero and points to your weight. At the same time, by Newton's third law the scale pushes up on you with a force equal to your weight, as shown in **Figure 19A.** This force balances the downward pull of gravity on you.

Free Fall and Weightlessness Now suppose you were standing on a scale in an elevator that is falling, as shown in **Figure 19B.** A falling object is in free fall when the only force acting on the object is gravity. Inside the free-falling elevator, you and the scale are both in free fall. Because the only force acting on you is gravity, the scale no longer is pushing up on you. According to Newton's third law, you no longer push down on the scale. So the scale pointer stays at zero and you seem to be weightless. **Weightlessness** is the condition that occurs in free fall when the weight of an object seems to be zero.

However, you are not really weightless in free fall because Earth is still pulling down on you. With nothing to push up on you, such as your chair, you would have no sensation of weight.

Figure 20
These oranges seem to be floating because they are falling around Earth at the same speed as the space shuttle and the astronauts. As a result, they don't seem to be moving relative to the astronauts in the cabin.

Weightlessness in Orbit To understand how objects move in the orbiting space shuttle, imagine you were holding a ball in the free-falling elevator. If you let the ball go, the position of the ball relative to you and the elevator wouldn't change, because you, the ball, and the elevator are moving at the same speed.

However, suppose you give the ball a gentle push downward. While you are pushing the ball, this downward force adds to the downward force of gravity. According to Newton's second law, the acceleration of the ball increases. So while you are pushing, the acceleration of the ball is greater than the acceleration of both you and the elevator. This causes the ball to speed up relative to you and the elevator. After it speeds up, it continues moving faster than you and the elevator, and it drifts downward until it hits the elevator floor.

When the space shuttle orbits Earth, the shuttle and all the objects in it are in free fall. They are falling in a curved path around Earth, instead of falling straight downward. As a result, objects in the shuttle appear to be weightless, as shown in **Figure 20**. A small push causes an object to drift away, just as a small downward push on the ball in the free-falling elevator caused it to drift to the floor.

Section 3 Assessment

1. You push a skateboard with a force of 6 N. If your mass is 60 kg, what is the force that the skateboard exerts on you?

2. You jump from a boat to a pier. As you move forward the boat moves backward. Explain.

3. What are the action and reaction forces when a hammer hits a nail?

4. Suppose you and a child that has half your mass are on skates. If the child gives you a push, who will have the greater acceleration? By how much?

5. **Thinking Critically** Suppose you are walking down the aisle of an airliner in flight. Use Newton's third law to describe the effect of your walk on the motion of the airliner.

Skill Builder Activities

6. **Solving One-Step Equations** A person standing on a canoe throws a cement block over the side. The action force on the cement block is 60 N. The reaction force is on the person and canoe. Their total mass is 100 kg. What is their acceleration? **For more help, refer to the** Math Skill Handbook.

7. **Communicating** Some people have trouble understanding Newton's third law. They reason, "If every action has an equal and opposite reaction, nothing ever will move." Explain why objects still can move. (Consider whether the forces act on the same or different objects.) **For more help, refer to the** Science Skill Handbook.

Balloon Races

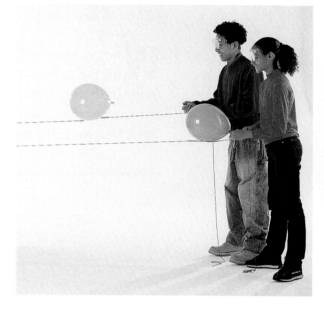

A balloon and a rocket lifting off the launch pad have something in common. Both use Newton's third law. In this experiment, you will compare different balloon rocket designs. The balloon rocket is powered by escaping air, and its motion is determined by Newton's first, second, and third laws.

What You'll Investigate
How do Newton's laws explain the motion of different balloon rockets?

Materials
balloons of different sizes and shapes
drinking straws
string
tape
meterstick
stopwatch*
*clock
*Alternate materials

Safety Precautions

Goals
■ **Measure** the speed of a balloon rocket.
■ **Describe** how Newton's laws explain a rocket's motion.

Procedure
1. Run a string across the classroom to make a rocket path. Leave one end loose so you can place the rockets on the string easily.
2. Make a balloon rocket according to the diagram. Don't tie the balloon. Let it run down the track. Measure the distance it travels and the time it takes.
3. Repeat step 2 with different balloons.

Conclude and Apply
1. **Compare and contrast** the distances traveled. Which rocket went the greatest distance?
2. **Calculate** the average speed for each rocket. Compare and contrast them. Which rocket has the greatest average speed?
3. **Infer** which aspects of these rockets made them travel far or fast.
4. **Draw** a diagram showing all the forces acting on a balloon rocket.
5. Use Newton's laws of motion to explain the motion of a balloon rocket from launch until it comes to a stop.

Communicating Your Data
Discuss with classmates which balloon rocket traveled the farthest. Why? **For more help, refer to the** Science Skill Handbook.

Activity
Design Your Own Experiment

Modeling Motion in Two Directions

When you move a computer mouse across a mouse pad, how does the rolling ball tell the computer cursor to move in the direction that you push the mouse? Inside the housing for the mouse's ball are two or more rollers that the ball rubs against as you move the mouse. They measure up-and-down and back-and-forth motions. What happens to the rollers when you move diagonally and at different angles?

Recognize the Problem

Can you move a golf ball from one point to another using forces in only two directions?

Form a Hypothesis

How can you combine forces to move in a straight line, along a diagonal, or around corners? Place a golf ball on something that will slide, such as a plastic lid. The plastic lid is called a skid. Lay out a course to follow on the floor. Write a plan for moving your golf ball along the path without having the golf ball roll away.

Possible Materials
masking tape
stopwatch*
*watch or clock with a second hand
meterstick*
*metric tape measure
spring scales marked in newtons (2)
plastic lid
golf ball*
*tennis ball
*Alternate materials

Goals
■ **Move** the skid across the ground using two forces.
■ **Measure** how fast the skid can be moved.
■ **Determine** how smoothly the direction can be changed.

Safety Precautions

Test Your Hypothesis

Plan

1. Lay out a course that involves two directions, such as always moving forward or left.

2. Attach two spring scales to the skid. One always will pull straight forward. One always will pull to one side. You cannot turn the skid. If one scale is pulling toward the door of your classroom, it always must pull in that direction. (It can pull with zero force if needed, but it can't push.)

3. How will you handle movements along diagonals and turns?

4. How will you measure speed?

5. **Experiment** with your skid. How hard do you have to pull to counteract sliding friction at a given speed? How fast can you accelerate? Can you stop suddenly without spilling the golf ball, or do you need to slow down?

6. **Write** a plan for moving your golf ball along the course by pulling only forward or to one side. Be sure you understand your plan and have considered all the details.

Do

1. Make sure your teacher approves your plan before you start.

2. Move your golf ball along the path.

3. Modify your plan, if needed.

4. **Organize** your data so they can be used to run your course and write them in your Science Journal.

5. **Test** your results with a new route.

Analyze Your Data

1. What was the difference between the two routes? How did this affect the forces you needed to use on the golf ball?

2. How did you separate and control variables in this experiment?

3. Was your hypothesis supported? Explain.

Draw Conclusions

1. What happens when you combine two forces at right angles?

2. If you could pull on all four sides (front, back, left, right) of your skid, could you move anywhere along the floor? Make a hypothesis to explain your answer.

*C*ommunicating
Your Data

Compare your conclusions with those of other students in your class. **For more help, refer to the** Science Skill Handbook.

TIME

SCIENCE AND
Society

SCIENCE
ISSUES
THAT AFFECT
YOU!

Air Bag Safety

After complaints and injuries, air bags in cars are helping all passengers

The car in front of yours stops suddenly. Your mom slams on the brakes, but not fast enough. You hear the crunch of car against car and feel your seat belt grab you. You look up at your mom in the front seat. She's covered with, not blood, thank goodness, but with a big white cloth. You are both okay. Your seat belts and air bags worked perfectly.

Popcorn in the Dash

Air bags have saved more than a thousand lives since 1992. They are like having a giant popcorn kernel in the dashboard that pops and becomes many times its original size. But unlike popcorn, an air bag is triggered by impact, not heat. When the air bag sensor picks up the vibrations of a crash, a chemical reaction is started. The reaction produces a gas that expands in a split second, inflating a balloonlike bag to cushion the driver and possibly the front-seat passenger. The bag deflates just as quickly so it doesn't trap people in the car.

Newton and the Air Bag

When you're traveling in a car, you move with it at whatever speed it is going. According to Newton's first law, you are the object in motion, and you will continue in motion unless acted upon by a force, such as a car crash.

Unfortunately, a crash stops the car, but it doesn't stop you, at least, not right away. You continue moving forward if your car doesn't have air bags or if you haven't buckled your seat belt. You stop when you strike the inside of the car. You hit the dashboard or steering wheel while traveling at the speed of the car. When an air bag inflates, it becomes the force acting on the moving object—you—and it stops you more gently.

A test measures the speed at which an air bag deploys.

Car manufacturers perform safety tests using dummies and air bags. The dummy is wearing a seat belt to simulate a human driver.

Unexpected Impact

"Our biggest issue was that air bags were not only helpful but dangerous—small children were being harmed by air bags," notes Betsy Ancker-Johnson, a spokesperson for an automobile-maker. She was referring to the fact that air bags pop out with so much force that they have sometimes hurt or killed children and small adults. For this reason, children under the age of 12 should ride in the back seat only, with seat belts buckled. Small adults may have their air bags turned off. Car makers are developing "smart" air bags that will expand just enough to protect a passenger no matter what the size or weight.

CONNECTIONS Measure Draw a steering wheel on a paper plate. Ask classmates to hold it 26 cm in front of them. That's the length drivers should have between the chest and the wheel to make air bags safe. Use a tape measure to check. Inform adult drivers in your family about this safety distance.

SCIENCE
Online
For more information, visit
science.glencoe.com

Reviewing Main Ideas

Section 1 Newton's First Law

1. A force is a push or a pull.

2. The net force is the combination of all the forces acting on an object.

3. Newton's first law states that objects in motion tend to stay in motion and objects at rest tend to stay at rest unless acted upon by a net force. *Why don't objects in motion on Earth, like a soccer ball, stay in motion forever?*

4. Friction is a force that resists motion between surfaces that are touching each other.

Section 2 Newton's Second Law

1. Newton's second law states that an object acted upon by a net force will accelerate in the direction of this force.

2. The acceleration due to a net force is given by the equation $a = F_{net}/m$. *If a baseball bat hits a bowling ball, why doesn't the bowling ball accelerate as quickly as a baseball that is hit just as hard?*

3. The force of gravity between two objects depends on their masses and the distance between them.

4. In circular motion, a force pointing toward the center of the circle acts on an object.

Section 3 Newton's Third Law

1. According to Newton's third law, the forces two objects exert on each other are always equal but in opposite directions. *What are the action and reaction forces acting on the skaters below?*

2. Action and reaction forces don't cancel because they act on different objects.

3. Action and reaction forces are involved in actions such as walking and jumping. Objects in orbit appear to be weightless because they are in free fall around Earth.

FOLDABLES
Reading & Study Skills

After You Read

Use the information in your Foldable to help you think of concrete examples for each law of motion. Write them under the tabs.

Visualizing Main Ideas

Fill in the following concept map on Newtons's laws of motion.

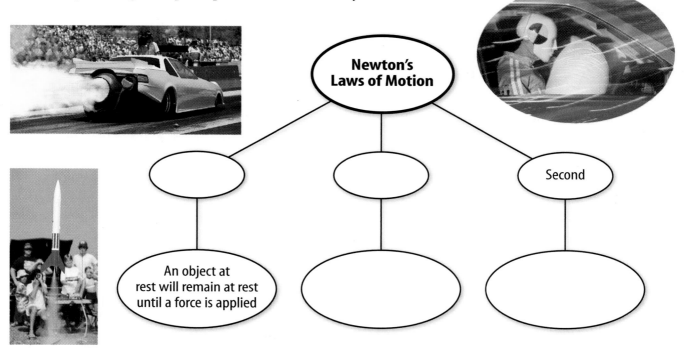

Newton's Laws of Motion

An object at rest will remain at rest until a force is applied

Second

Vocabulary Review

Vocabulary Words

a. balanced forces
b. force
c. friction
d. net force
e. Newton's first law of motion
f. Newton's second law of motion
g. Newton's third law of motion
h. unbalanced forces
i. weight
j. weightlessness

Using Vocabulary

Explain the differences between the terms in the following sets.

1. force, inertia, weight
2. Newton's first law of motion, Newton's third law of motion
3. friction, force
4. net force, balanced forces
5. weight, weightlessness
6. balanced forces, unbalanced forces
7. friction, weight
8. Newton's first law of motion, Newton's second law of motion
9. friction, unbalanced force
10. net force, Newton's third law of motion

Chapter 2 Assessment

Checking Concepts

Choose the word or phrase that best answers the question.

1. Which of the following changes when an unbalanced force acts on an object?
   - **A)** mass
   - **B)** motion
   - **C)** inertia
   - **D)** weight

2. Which of the following is the force that slows a book sliding on a table?
   - **A)** gravity
   - **B)** static friction
   - **C)** sliding friction
   - **D)** inertia

3. What combination of units is equivalent to the newton?
   - **A)** m/s^2
   - **B)** $kg \cdot m/s$
   - **C)** $kg \cdot m/s^2$
   - **D)** kg/m

4. What is a push or pull a definition of?
   - **A)** force
   - **B)** momentum
   - **C)** acceleration
   - **D)** inertia

5. What is the type of friction that is important to walking?
   - **A)** static friction
   - **B)** sliding friction
   - **C)** rolling friction
   - **D)** air resistance

6. An object is accelerated by a net force in which direction?
   - **A)** at an angle to the force
   - **B)** in the direction of the force
   - **C)** in the direction opposite to the force
   - **D)** Any of these is possible.

7. If you exert a net force of 8 N on a 2-kg object, what will its acceleration be?
   - **A)** 4 m/s^2
   - **B)** 6 m/s^2
   - **C)** 12 m/s^2
   - **D)** 16 m/s^2

8. You are riding on a bike. In which of the following situations are the forces acting on the bike balanced?
   - **A)** You pedal to speed up.
   - **B)** You turn at constant speed.
   - **C)** You coast to slow down.
   - **D)** You pedal at constant speed.

9. Which of the following has no direction?
   - **A)** force
   - **B)** acceleration
   - **C)** weight
   - **D)** mass

10. You push against a wall with a force of 5 N. What is the force the wall exerts on your hands?
    - **A)** 0 N
    - **B)** 2.5 N
    - **C)** 5 N
    - **D)** 10 N

Thinking Critically

11. A baseball is pitched east at a speed of 40 km/h. The batter hits it west at a speed of 40 km/h. Did the ball accelerate? Explain.

12. Frequently, the pair of forces acting between two objects are not noticed because one of the objects is Earth. Explain why the force acting on Earth isn't noticed.

13. A car is parked on a hill. The driver starts the car, accelerates until the car is driving at constant speed, drives at constant speed, and then brakes to put the brake pads in contact with the spinning wheels. Explain how static friction, sliding friction, rolling friction, and air resistance are acting on the car.

14. You hit a hockey puck and it slides across the ice at nearly a constant speed. Is a force keeping it in motion? Explain.

15. Newton's third law describes the forces between two colliding objects. Use this connection to explain the forces acting when you kick a soccer ball.

Developing Skills

16. **Recognizing Cause and Effect** Use Newton's third law to explain how a rocket accelerates upon takeoff.

17. Prediciting Two balls of the same size and shape are dropped from a helicopter. One ball has twice the mass of the other ball. On which ball will the force of air resistance be greater when terminal velocity is reached?

18. Interpreting Scientific Illustrations Is the force on the box balanced? Explain.

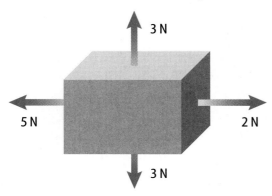

19. Solving One-Step Equations A 0.4-kg object accelerates at 2 m/s². Find the force.

Performance Assessment

20. Oral Presentation Research one of Newton's laws of motion and compose an oral presentation. Provide examples of the law. You might want to use a visual aid.

21. Writing in Science Create an experiment that deals with Newton's laws of motion. Document it using the following subject heads: *Title of Experiment, Partners' Names, Hypothesis, Materials, Procedures, Data, Results,* and *Conclusion.*

TECHNOLOGY

Go to the Glencoe Science Web site at **science.glencoe.com** or use the **Glencoe Science CD-ROM** for additional chapter assessment.

THE PRINCETON REVIEW Test Practice

The following diagram shows an experiment in which data were collected about falling objects.

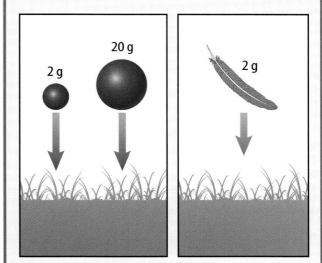

Study the diagram above and then answer the following questions about the experiment.

1. Which of these is the most likely hypothesis for the experiment depicted in the box on the left above?
  A) The more mass an object has, the faster it will travel.
  B) The less mass an object has, the faster it will travel.
  C) Objects with less mass travel faster than objects with more mass.
  D) Objects of different masses can still travel at the same speed.

2. The feather is traveling at a different speed than the balls because the feather _____ .
  F) has more gravity acting upon it
  G) has less gravity acting upon it
  H) has more friction acting upon it
  J) has less friction acting upon it

Forces and Fluids

A piece of metal such as a fork or a spoon sinks to the bottom of a kitchen sink filled with water. This giant supertanker sailing in the waters off the coast of Alaska is made of a lot more metal than a spoon, yet it floats, even when it is filled with thousands of barrels of oil. Why do some objects float and some objects sink? In this chapter, you will learn how forces are transmitted through fluids, why objects float, and how fluids can be used to perform work.

What do you think?

Science Journal Look at the picture below with a classmate. Discuss what you think this might be or what is happening. Here's a hint: *Even if you had these, you still couldn't fly.* Write your answer or best guess in your Science Journal.

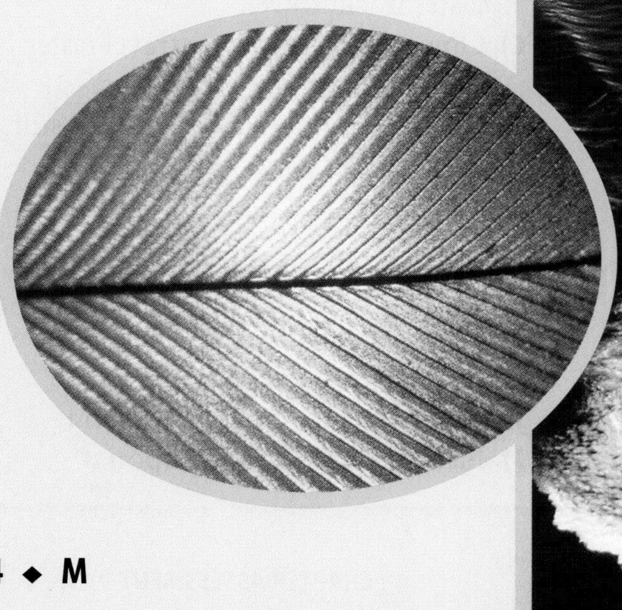

EXPLORE ACTIVITY

When you are lying down, something is pushing down on you with a force equal to the weight of several small cars. What is the substance that is applying all this pressure on you? Air. You use air pressure every time you ride on automobile or bicycle tires, use a vacuum cleaner, or drink from a straw. Explore the effects of air pressure during this activity.

Experiment with air pressure

1. Suck water into a straw. Try to keep the straw completely filled with water.
2. Quickly cap the top of the straw with your finger and observe what happens to the water.
3. Release your finger from the top of the straw for an instant and replace it as quickly as possible. Observe what happens to the water.
4. Release your finger from the top of the straw and observe.

Observe

Write a paragraph in your Science Journal describing your observation of the water in the straw. Make a hypothesis as to what force keeps water from falling out of the straw when you cap the top of the straw.

Before You Read

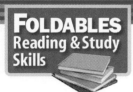

FOLDABLES
Reading & Study Skills

Making a Venn Diagram Study Fold Make the following Foldable to compare and contrast the characteristics of two fluids—liquids and gases.

1. Place a sheet of paper in front of you so the long side is at the top. Fold the paper in half from top to bottom.
2. Fold both sides in. Unfold the paper so three sections show.
3. Through the top thickness of paper, cut along each of the fold lines to the fold, forming three tabs.
4. Draw ovals and label the tabs *Liquids, Fluids,* and *Gases* as shown.
5. As you read the chapter, write information about fluids under the middle tab.

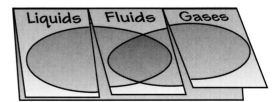

Pressure

As You Read

What You'll Learn
- **Define and calculate** pressure.
- **Model** how pressure varies in a fluid.

Vocabulary
pressure
fluid

Why It's Important
Some of the processes that help keep you alive depend on differences in pressure.

What is pressure?

What happens when you walk in deep, soft snow or dry sand? Your feet sink into the snow or sand and walking can be difficult. If you rode a bicycle with narrow tires over these surfaces, the tires would sink even deeper than your feet.

How deep you sink depends on your weight as well as the area over which you make contact with the sand or snow. Like the person in **Figure 1,** when you stand on two feet, you make contact with the sand over the area covered by your feet. However, if you were to stand on a large piece of wood, your weight would be distributed over the area covered by the wood.

In both cases, your weight exerted a downward force on the sand. What changed was the area of contact between you and the sand. By changing the area of contact, you changed the pressure you exerted on the sand due to your weight. **Pressure** is the force per unit area that is applied on the surface of an object. When you stood on the board, the area of contact increased, so that the same force was applied over a larger area. As a result, the pressure that was exerted on the sand decreased and you didn't sink as deep.

Figure 1
When your weight is distributed over a larger area, the pressure you exert on the sand decreases.

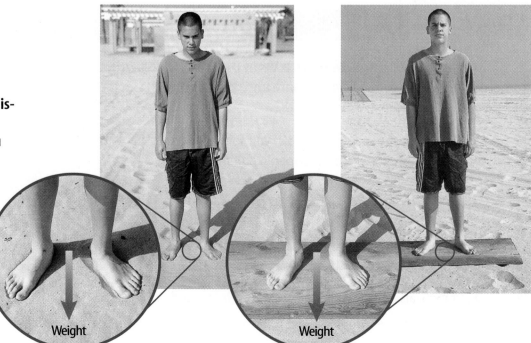

Calculating Pressure What would happen to the pressure exerted by your feet if your weight increased? You might expect that you would sink deeper in the sand, so the pressure also would increase. Pressure increases if the force applied increases, and decreases if the area of contact increases. Pressure can be calculated from this formula.

$$\text{Pressure (Pa)} = \frac{\text{force (N)}}{\text{area (m}^2)}$$

$$P = F/A$$

The unit of pressure in the SI system is the pascal, abbreviated Pa. One pascal is equal to a force of 1 N applied over an area of 1 m², or $1 \text{ Pa} = 1 \text{ N/m}^2$. The weight of a dollar bill resting completely flat on a table exerts a pressure of about 1 Pa on the table. Because 1 Pa is a small unit of pressure, pressure sometimes is expressed in units of kPa, which is 1,000 Pa.

SCIENCE *Online*

Research Visit the Glencoe Science Web site at **science.glencoe.com** for information about the history and use of snowshoes. These devices have been used for centuries in cold, snowy climates. Communicate to your class what you learn.

Math Skills Activity

Calculating Pressure

Example Problem

A water glass sitting on a table weighs 4 N. The bottom of the water glass has a surface area of 0.003 m². Calculate the pressure the water glass exerts on the table.

Solution

1 *This is what you know:*   force: $F = 4$ N
                                               area: $A = 0.003$ m²

2 *This is what you need to find:*   pressure: P

3 *This is the formula you need to use:* $P = F/A$

4 *Substitute the known values:* $P = (4 \text{ N})/(0.003 \text{ m}^2)$
                                                    $= 1{,}333 \text{ N/m}^2 = 1{,}333$ Pa

Check your answer by multiplying it by the given area. Did you calculate the force that was given?

Practice Problem

A student weighs 600 N. The student's shoes are in contact with the floor over a surface area of 0.012 m². Calculate the pressure exerted by the student on the floor.

For more help, refer to the Math Skill Handbook.

Interpreting Footprints

Procedure

1. Go outside to **an area of dirt, sand, or snow** where you can make footprints. Smooth the surface.
2. Make tracks in several different ways. Possible choices include walking forward, walking backward, running, jumping a short or long distance, walking carrying a load, and tiptoeing.

Analysis

1. Measure the depth of each type of track at two points: the ball of the foot and the heel. Compare the depths of the different tracks.
2. The depth of the track corresponds to the pressure on the ground. In your **Science Journal,** explain how different means of motion put pressure on different parts of the sole.
3. Finally, have one person make a track while the other looks away. Then have the second person determine what the motion was.

Pressure and Weight To calculate the pressure that is exerted on a surface, you need to know the force and the area over which it is applied. Sometimes the force that is exerted is the weight of an object, such as when you are standing on sand, snow, or a floor. Suppose you are holding a 2-kg book in the palm of your hand. To find out how much pressure is being exerted on your hand, you first must know the force that the book is exerting on your hand—its weight.

$$\text{Weight} = \text{mass} \times \text{acceleration due to gravity}$$
$$W = (2 \text{ kg}) \times (9.8 \text{ m/s}^2)$$
$$W = 19.6 \text{ N}$$

If the area of contact between your hand and the book is 0.003 m², the pressure that is exerted on your hand by the book is:

$$P = F/A$$
$$P = (19.6 \text{ N})/(0.003 \text{ m}^2)$$
$$P = 6{,}533 \text{ Pa} = 6.53 \text{ kPa}$$

Pressure and Area One way to change the pressure that is exerted on an object is to change the area over which the force is applied. Imagine trying to drive a nail into a piece of wood, as shown in **Figure 2.** Why is the tip of a nail pointed instead of flat? When you hit the nail with a hammer, the force you apply is transmitted through the nail from the head to the tip. The tip of the nail comes to a point and is in contact with the wood over a small area. Because the contact area is so small, the pressure that is exerted by the nail on the wood is large—large enough to push the wood fibers apart. This allows the nail to move downward into the wood.

✔ **Reading Check** *How does changing area affect pressure?*

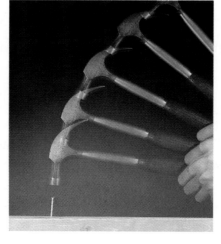
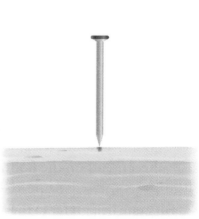

Figure 2
The force applied to the head of the nail by the hammer is the same as the force that the tip of the nail applies to the wood. However, because the area of the tip is small, the pressure applied to the wood is large.

The water can be squeezed from its container into the pan.

Water flows from the jug to the canteen, taking the shape of the canteen.

The gas completely fills the propane tank. After it is attached to the stove, the gas can flow to be used as fuel.

Fluids

What do the substances in **Figure 3** have in common? Each takes the shape of its container and can flow from one place to another. A **fluid** is any substance that has no definite shape and has the ability to flow. You might think of a fluid as being a liquid, such as water or motor oil. But gases are also fluids. When you are outside on a windy day, you can feel the air flowing past you. Because air can flow and has no definite shape, air is a fluid. Gases, liquids, and the state of matter called plasma, which is found in the Sun and other stars, are fluids and can flow.

Pressure in a Fluid

Suppose you placed an empty glass on a table. The weight of the glass exerts pressure on the table. If you fill the glass with water, the weight of the water and glass together exert a force on the table. So the pressure exerted on the table increases.

Because the water has weight, the water itself also exerts pressure on the bottom of the glass. This pressure is the weight of the water divided by the area of the glass bottom. If you pour more water into the glass, the height of the water in the glass increases and the weight of the water increases. As a result, the pressure exerted by the water increases.

Figure 3
Fluids all have the ability to flow and take the shape of their containers. *What are some other examples of fluids?*

Astronomy
INTEGRATION

The Sun is a star with a core temperature of about 16 million°C. At this temperature, the particles in the Sun move at tremendous speeds, crashing into each other in violent collisions that tear atoms apart. The Sun is made of a gas of electrically charged particles. A gas made of electrically charged particles is a plasma. So the Sun is made of a plasma.

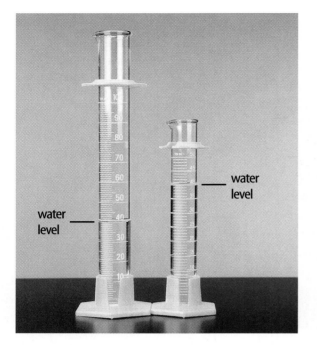

Figure 4
Even though each graduated cylinder contains the same volume of water, the pressure exerted by the higher column of water is greater.

Pressure and Fluid Height
Suppose you poured the same amount of water into a small and a large graduated cylinder, as shown in **Figure 4.** Notice that the height of the water in the small cylinder is greater than in the large cylinder. Is the water pressure the same at the bottom of each cylinder? The weight of the water in each cylinder is the same, but the contact area at the bottom of the small cylinder is smaller. Therefore, the pressure is greater at the bottom of the small cylinder.

The height of the water can increase if more water is added to a container or if the same amount of water is added to a narrower container. In either case, when the height of the fluid is greater, the pressure at the bottom of the container is greater. This is always true for any fluid or any container. The greater the height of a fluid above a surface, the greater the pressure exerted by the fluid on that surface. The pressure exerted at the bottom of a container doesn't depend on the shape of the container, but only on the height of the fluid above the bottom, as **Figure 5** shows.

Going Deeper
If you swim underwater, you might notice that you can feel pressure in your ears. As you go deeper, you can feel this pressure increase. This pressure is exerted by the weight of the water above you. As you go deeper in a fluid, the height of the fluid above you increases. As the height of the fluid above you increases, the weight of the fluid above you also increases. As a result, the pressure exerted by the fluid increases with depth.

Figure 5
Pressure depends only on the height of the fluid above a surface, not on the shape of the container. The pressure at the bottom of each section of the tube is the same.

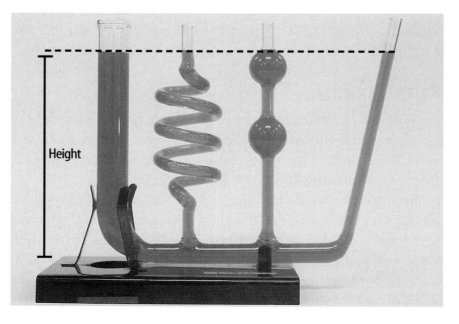

Height

Pressure in All Directions If the pressure that is exerted by a fluid is due to the weight of the fluid, is the pressure in a fluid exerted only downward? **Figure 6** shows a small, solid cube in a fluid. The fluid exerts a pressure on each side of this cube, not just on the top. The pressure on each side is perpendicular to the surface, and the amount of pressure depends only on the depth in the fluid. As shown in **Figure 6,** this is true for any object in a fluid, no matter how complicated the shape. The pressure at any point on the object is perpendicular to the surface of the object at that point.

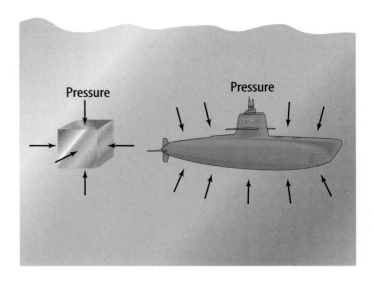

✔ **Reading Check** *In what direction is pressure by a fluid exerted?*

Figure 6
The pressure on all objects in a fluid is exerted on all sides, perpendicular to the surface of the object, no matter what its shape.

Atmospheric Pressure

Even though you don't feel it, you are surrounded by a fluid that exerts pressure on you constantly. That fluid is the atmosphere. The atmosphere at Earth's surface is only about one-thousandth as dense as water. However, the thickness of the atmosphere is large enough to exert a large pressure on objects at Earth's surface. For example, look at **Figure 7.** When you are sitting down, the force pushing down on your body due to atmospheric pressure can be equal to the weight of several small cars. Atmospheric pressure is approximately 100,000 Pa at sea level. This means that the weight of Earth's atmosphere exerts about 100,000 N of force over every square meter on Earth.

Why doesn't this pressure cause you to be crushed? Your body is filled with fluids such as blood that also exert pressure. The pressure exerted outward by the fluids inside your body balances the pressure exerted by the atmosphere.

Figure 7
Atmospheric pressure on your body is a result of the weight of the atmosphere exerting force on your body. *Why don't you feel the pressure exerted by the atmosphere?*

Going Higher As you go higher in the atmosphere, atmospheric pressure decreases as the amount of air above you decreases. The same is true in an ocean, lake, or pond. The water pressure is highest at the ocean floor and decreases as you go upward. The changes in pressure at varying heights and depths are illustrated in **Figure 8.**

Figure 8

No matter where you are on Earth, you're under pressure. Air and water have weight and therefore exert pressure on your body. The amount of pressure depends on your location above or below sea level and how much air or water—or both—are exerting force on you.

8,000 m

7,000 m

6,000 m

5,000 m

4,000 m

3,000 m

2,000 m

1,000 m

▲ HIGH ELEVATION With increasing elevation, the amount of air above you decreases, and so does air pressure. At the 8,850-m summit of Mt. Everest, air pressure is a mere 33 kPa—about one third of the air pressure at sea level.

0 m

▲ SEA LEVEL Air pressure is pressure exerted by the weight of the atmosphere above you. At sea level the atmosphere exerts a pressure of about 100,000 N on every square meter of area. Called one atmosphere (atm), this pressure is also equal to 100 kPa.

1,000 m

▶ REEF LEVEL When you descend below the sea surface, pressure increases by about 1 atm every 10 meters. At 20 meters depth, you'd experience 2 atm of water pressure and 1 atm of air pressure, a total of 3 atm of pressure on your body.

2,000 m

3,000 m

4,000 m

5,000 m

6,000 m

7,000 m

▶ VERY LOW ELEVATION The deeper you dive, the greater the pressure. The water pressure on a submersible at a depth of 2,200 m is about 220 times greater than atmospheric pressure at sea level.

8,000 m

9,000 m

Barometer An instrument called a barometer is used to measure atmospheric pressure. A barometer has something in common with a drinking straw. When you drink through a straw, it seems like you pull your drink up through the straw. But actually, atmospheric pressure pushes your drink up the straw. By removing air from the straw, you reduce the air pressure in the straw. Meanwhile, the atmosphere is pushing down on the surface of your drink. When you pull the air from the straw, the pressure in the straw is less than the pressure pushing down on the liquid, so atmospheric pressure pushes the drink up the straw.

One type of barometer works in a similar way, as shown in **Figure 9.** The space at the top of the tube is a vacuum. Atmospheric pressure pushes liquid up a tube. The liquid reaches a height where the pressure at the bottom of the column of liquid balances the pressure of the atmosphere. As the atmospheric pressure changes, the force pushing on the surface of the liquid changes. As a result, the height of the liquid in the tube increases as the atmospheric pressure increases.

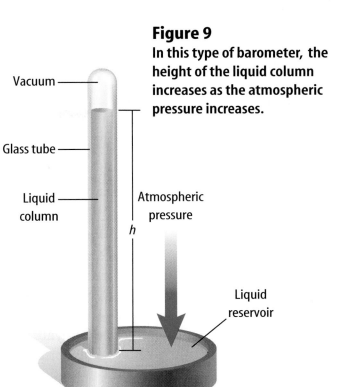

Figure 9
In this type of barometer, the height of the liquid column increases as the atmospheric pressure increases.

Vacuum
Glass tube
Liquid column
Atmospheric pressure
h
Liquid reservoir

Section 1 Assessment

1. Define pressure and give an example of an item that exerts pressure.
2. Calculate the pressure that a 50-kg woman wearing high heels exerts on a floor. Both shoes are in contact with the floor over a total area of 0.007 m².
3. Why is it hard to walk across soft ground in spike heels?
4. As the depth in a fluid changes, how does the pressure change? Explain.
5. **Think Critically** A balloon is partially inflated and then carried up a high mountain. Will the balloon expand, shrink, or remain the same? Explain.

Skill Builder Activities

6. **Classifying** Classify the following as fluids or solids: *warm butter, paper, liquid nitrogen, neon gas, baked clay,* and *soil.* **For more help, refer to the** Science Skill Handbook.
7. **Using an Electronic Spreadsheet** Create a spreadsheet from the following barometric pressures in kPa that were recorded on December 5: *Dallas, 88.5; Denver, 88.8; San Diego, 88.2; Columbus, 89.0; New York City, 87.8; Bangor, 87.0;* and *Seattle, 89.1.* Use the spreadsheet to calculate the average pressure from these data. **For more help, refer to the** Technology Skill Handbook.

Why do objects float?

What You'll Learn

- **Explain** how the pressure in a fluid produces a buoyant force.
- **Define** density.
- **Explain** floating and sinking using Archimedes' principle.

Vocabulary

buoyant force
Archimedes' principle
density

Why It's Important

Knowing how fluids exert forces helps you understand how boats can float.

The Buoyant Force

Can you float? Think about the forces that are acting on you as you float motionless on the surface of a pool or lake. You are not moving, so according to Newton's second law of motion, the forces on you must be balanced. Earth's gravity is pulling you downward, so an upward force must be balancing your weight, as shown in **Figure 10.** This force is called the buoyant force. The **buoyant force** is an upward force that is exerted by a fluid on any object in the fluid.

What causes the buoyant force?

The buoyant force is caused by the pressure that is exerted by a fluid on an object in the fluid. **Figure 11** shows a cube-shaped object submerged in a glass of water. The water exerts pressure everywhere over the surface of the object. Recall that the pressure exerted by a fluid has two properties. One is that the direction of the pressure on a surface is always perpendicular to the surface. The other is that the pressure exerted by a fluid increases as you go deeper into the fluid.

Figure 10
When you float, the forces on you are balanced. Gravity pulls you downward and is balanced by the buoyant force pushing you upward.

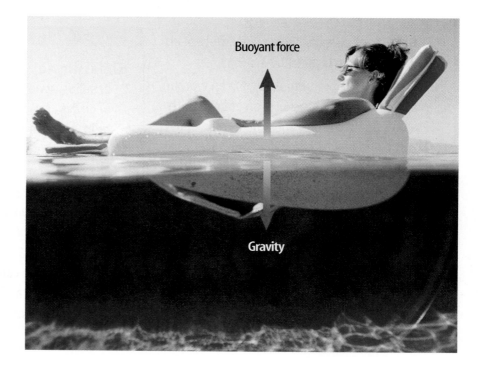

Buoyant force

Gravity

Buoyant Force and Unbalanced Pressure The pressure that is exerted by the water on the cube is shown in **Figure 11.** The bottom of the cube is deeper in the water. Therefore, the pressure that is exerted by the water at the bottom of the cube is greater than it is at the top of the cube. The higher pressure near the bottom means that the water exerts an upward force on the bottom of the cube that is greater than the downward force that is exerted at the top of the cube. As a result, the force that is exerted on the cube due to water pressure is not balanced, and a net upward force is acting on the cube due to the pressure of the water. This upward force is the buoyant force. A buoyant force acts on all objects that are placed in a fluid, whether they are floating or sinking.

✔ Reading Check *When does the buoyant force act on an object?*

Sinking and Floating

If you drop a stone into a pool of water, it sinks. But if you toss a twig on the water, it floats. An upward buoyant force acts on the twig and the stone, so why does one float and one sink?

The buoyant force pushes an object in a fluid upward, but gravity pulls the object downward. If the weight of the object is greater than the buoyant force, the net force on the object is downward and it sinks. If the buoyant force is equal to the object's weight, the forces are balanced and the object floats. As shown in **Figure 12,** the fish floats because the buoyant force on it balances its weight. The rocks sink because the buoyant force acting on them is not large enough to balance their weight.

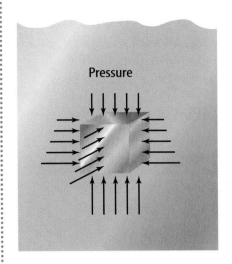

Figure 11
The pressure exerted on the bottom of the cube is greater than the pressure on the top. The fluid exerts a net upward force on the cube.

Figure 12
The weight of a rock is more than the buoyant force exerted by the water, so it sinks to the bottom. *Why do the fish float?*

Figure 13
A piece of aluminum foil sinks when it is folded up so that its surface area is small. The surface area of the foil increases when it is unfolded. Thus, the buoyant force is larger, and it floats.

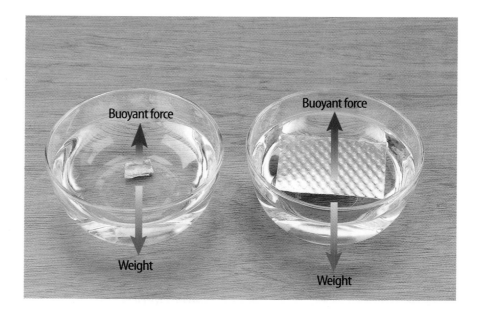

Changing the Buoyant Force

Whether an object sinks or floats depends on whether the buoyant force is smaller than its weight. The weight of an object depends only on the object's mass, which is the amount of matter the object contains. The weight does not change if the shape of the object changes. A piece of modeling clay contains the same amount of matter whether it's squeezed into a ball or pressed flat.

Buoyant Force and Shape Buoyant force does depend on the shape of the object. The fluid exerts upward pressure on the entire lower surface of the object that is in contact with the fluid. If this surface is made larger, then more upward pressure is exerted on the object and the buoyant force is greater. **Figure 13** shows how a piece of aluminum can be made to float.

If the aluminum is crumpled, the buoyant force is less than the weight, so the aluminum sinks. When the aluminum is flattened into a thin, curved sheet, the buoyant force is large enough that the sheet floats. This is how large, metal ships, like the one in **Figure 14,** are able to float. The metal is formed into a curved sheet that is the hull of the ship. The contact area of the hull with the water is much greater than if the metal were a solid block. As a result, the buoyant force on the hull is greater than it would be on a metal block.

Figure 14
Even though it is made of metal, the hull of this oil tanker has a large surface area in contact with the water. As a result, the buoyant force is large enough that the ship floats.

The Buoyant Force Doesn't Change with Depth

Suppose you drop a steel cube into the ocean. You might think that the cube would sink only to a depth where the buoyant force on the cube balances its weight. However, the steel sinks to the bottom, no matter how deep the ocean is.

The buoyant force on the cube is the difference between the downward force due to the water pressure on the top of the cube and the upward force due to water pressure on the bottom of the cube. **Figure 15** shows that when the cube is deeper, the pressure on the top surface increases, but the pressure on the bottom surface also increases by the same amount. As a result, the difference between the forces on the top and bottom surfaces is the same, no matter how deep the cube is submerged. The buoyant force on the submerged cube is the same at any depth.

Archimedes' Principle

A way of determining the buoyant force was given by the ancient Greek mathematician Archimedes(ar kuh MEE deez) more than 2,200 years ago. According to **Archimedes' principle,** the buoyant force on an object is equal to the weight of the fluid it displaces.

To understand Archimedes' principle, think about what happens if you drop an ice cube in a glass of water that's filled to the top. The ice cube takes the place of some of the water and causes this water to overflow, as shown in **Figure 16.** Another way to say this is that the ice cube displaced water that was in the glass.

Suppose you caught all the overflow water and weighed it. According to Archimedes' principle, the weight of the overflow, or displaced water, would be equal to the buoyant force on the ice cube. Because the ice cube is floating, the buoyant force is balanced by the weight of the ice cube. So the weight of the water that is displaced, or the buoyant force, is equal to the weight of the ice cube.

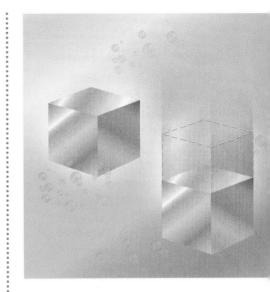

Figure 15
Because the cube on the right is deeper, the pressure on its upper surface is increased due to the weight of the water inside the dashed lines. The pressure on the bottom surface also increases by this amount.

A

B

Figure 16
The buoyant force exerted on this ice cube is equal to the weight of the water displaced by the ice cube.

Icebergs are chunks of ice that break off glaciers and ice shelves and float in the ocean. Icebergs float because ice is less dense than water. Icebergs vary in size, but one of the largest ever reported was about the size of the state of Connecticut. Typically, only about ten percent of the total volume of the iceberg is above water. The remaining 90 percent is submerged.

Density Archimedes' principle leads to a way of determining whether an object placed in a fluid will float or sink. The answer depends on comparing the density of the fluid and the density of the object. The **density** of a fluid or an object is the mass of the object divided by the volume it occupies. Density can be calculated by the following formula:

$$\text{Density} = \text{mass/volume}$$

For example, water has a density of 1.0 g/cm^3 and copper has a density of 9.0 g/cm^3. The density of copper is 9.0 times greater than the density of water. The mass of any volume of a substance can be calculated by multiplying both sides of the above equation by volume.

$$\text{Mass} = \text{density} \times \text{volume}$$

For example, a volume of copper has a mass 9.0 times larger than an equal volume of water, because the density of copper is 9.0 times greater than the density of water.

Problem-Solving Activity

Layering Liquids

The density of an object or substance determines whether it will sink or float in a fluid. Just like solid objects, liquids also have different densities. If you pour vegetable oil into water, the oil doesn't mix. Instead, because the density of oil is less than the density of water, the oil floats on top of the water.

Identifying the Problem

In science class, a student is presented with five unknown liquids and their densities. He measures the volume of each and organizes his data into the table at the right. He decides to experiment with these liquids by carefully pouring them, one at a time, into a graduated cylinder.

Liquid Density and Volume			
Liquid	Color	Density (g/cm³)	Volume (cm³)
A	red	2.40	32.0
B	blue	2.90	15.0
C	green	1.20	20.0
D	yellow	0.36	40.0
E	purple	0.78	19.0

Solving the Problem

1. Assuming the liquids don't mix with each other, draw a diagram and label the colors, illustrating how these liquids would look when poured into a graduated cylinder. If 30 cm^3 of water were added to the graduated cylinder, explain how your diagram would change.

2. Use the formula for density to calculate the mass of each of the unknown liquids in the chart.

Sinking and Density Suppose you place a copper block with a volume of 1,000 cm³ into a container of water. This block weighs about 88 N. As the block sinks, it displaces water, and an upward buoyant force acts on it. If the block is completely submerged, the volume of water it has displaced is 1,000 cm³—the same as its own volume. This is the maximum amount of water the block can displace. The weight of 1,000 cm³ of water is about 10 N, and this is the maximum buoyant force that can act on the block. This buoyant force is less than the weight of the copper, so the copper block continues to sink.

The copper block and the displaced water had the same volume. Because the copper block had a greater density, the mass of the copper block was greater than the mass of the displaced water. As a result, the copper block weighed more than the displaced water because its density was greater. Any material with a density that is greater than the density of water will weigh more than the water that it displaces, and it will sink. This is true for any object and any fluid. Any object that has a density greater than the density of the fluid it is placed in will sink.

Floating and Density Suppose you place a block of wood with a volume of 1,000 cm³ into a container of water. This block weighs about 7 N. The block starts to sink and displaces water. However, it stops sinking and floats before it is completely submerged, as shown in **Figure 17.** The density of the wood was less than the density of the water. So the wood was able to displace an amount of water equal to its weight before it was completely submerged. It stopped sinking after it had displaced about 700 cm³ of water. That much water has a weight of about 7 N, which is equal to the weight of the block. Any object with a density less than the fluid it is placed in will float.

Reading Check *How can you determine whether an object will float or sink?*

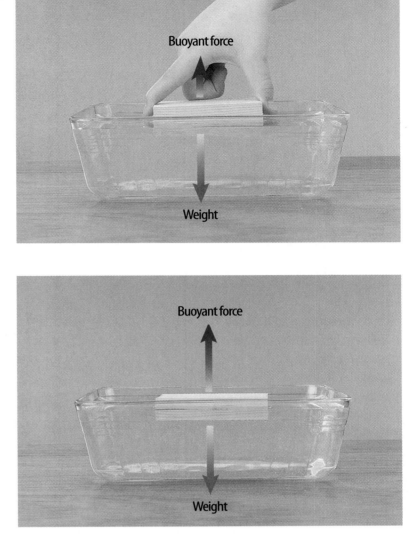

Figure 17
An object, such as this block of wood, will continue to sink in a fluid until it has displaced an amount of fluid that is equal to its mass. Then the buoyant force equals its weight.

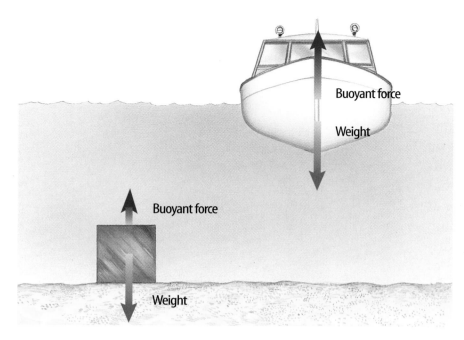

Buoyant force

Weight

Buoyant force

Weight

Figure 18
Even though the boat and the cube have the same mass, the boat displaces much more water because of its shape, reducing its overall density. Therefore the boat is able to float, but the cube sinks.

Boats

Archimedes' principle provides another way to understand why boats that are made of metal can float. Look at **Figure 18.** By making a piece of steel into a boat that occupies a large volume, more water is displaced by the boat than by the piece of steel. According to Archimedes' principle, increasing the weight of the water that is displaced increases the buoyant force. By making the volume of the boat large enough, enough water can be displaced so that the buoyant force is greater than the weight of the steel.

How does the density of the boat compare to the density of the piece of steel? The steel now surrounds a volume that is filled with air that has little mass. The mass of the boat is nearly the same as the mass of the steel, but the volume of the boat is much larger. As a result, the density of the boat is much less than the density of the steel. The boat floats when its volume becomes large enough that its density is less than the density of water.

Section ② Assessment

1. Does the buoyant force on a submerged object depend on the weight of the object? Explain your answer.

2. Explain what the term *density* means.

3. If an object has a density of 1.5 g/cm^3, will it float or sink in water? Explain.

4. Using Archimedes' principle, explain why a rowboat will float. If the boat fills with water, explain why it will sink.

5. **Think Critically** A submarine changes its mass by adding or removing seawater from tanks inside the sub. Explain how this can enable the sub to dive or rise to the surface.

Skill Builder Activities

6. **Recognizing Cause and Effect** Explain why a log floats on a river until it passes under a waterfall, where it sinks until it moves beyond the waterfall. Use the terms *buoyant force, density,* and *net force* in your explanation. **For more help, refer to the** Science Skill Handbook.

7. **Solving One-Step Equations** A ship displaces 80,000 L of water. One liter of water weighs 9.8 N. What is the upward force on the ship? **For more help, refer to the** Math Skill Handbook.

Activity

Measuring Buoyant Force

The total force on an object in a fluid is the difference between the object's weight and the buoyant force. In this activity, you will measure the buoyant force on an object and compare it to the weight of the water displaced.

What You'll Investigate

How is the buoyant force related to the weight of the water that an object displaces?

Materials

aluminum pan   graduated cylinder
spring scale   funnel
500-mL beaker   metal object

Goals

- **Measure** the buoyant force on an object.
- **Compare** the buoyant force to the weight of the water displaced by the object.

Safety Precautions

Procedure

1. Place the beaker in the aluminum pan and fill the beaker to the brim with water.

2. Hang the object from the spring scale and record its weight.

3. With the object hanging from the spring scale, completely submerge the object in the water. The object should not be touching the bottom or the sides of the beaker.

4. **Record** the reading on the spring scale while the object is in the water. Calculate the buoyant force by subtracting this reading from the object's weight.

5. Use the funnel to carefully pour the water from the pan into the graduated cylinder. Record the volume of this water in cm^3.

6. **Calculate** the weight of the water displaced by multiplying the volume of water by 0.0098 N.

Conclude and Apply

1. How did the total force on the object change when it was submerged in water?

2. **Compare** the weight of the water that is displaced with the buoyant force.

3. How would the buoyant force change if the object were submersed halfway in water?

Communicating Your Data

Make a poster of an empty ship, a heavily loaded ship, and an overloaded, sinking ship. Explain how Archimedes' principle applies in each case. **For more help, refer to the** Science Skill Handbook.

Doing Work with Fluids

As You Read

What You'll Learn

- **Explain** how forces are transmitted through fluids.
- **Describe** how a hydraulic system increases force.
- **Describe** Bernoulli's principle.

Vocabulary
Pascal's principle
hydraulic system
Bernoulli's principle

Why It's Important
Fluids can exert forces that lift heavy objects and enable aircraft to fly.

Using Fluid Forces

You might have watched a hydraulic lift raise a car off the ground. It might surprise you to learn that the force pushing the car upward is being exerted by a fluid. When a huge jetliner soars through the air, a fluid exerts the force that holds it up. Fluids at rest and fluids in motion can be made to exert forces that do useful work, such as pumping water from a well, making cars stop, and carrying people long distances through the air. How are these forces produced by fluids?

Pushing on a Fluid The pressure in a fluid can be increased by pushing on the fluid. Suppose a watertight, movable cover, or piston, is sitting on top of a column of fluid in a container. If you push on the piston, the fluid can't escape past the piston, so the height of the fluid in the container doesn't change. As a result, the piston doesn't move. But now the force exerted on the bottom of the container is the weight of the fluid plus the force pushing the piston down. Because the force exerted by the fluid at the bottom of the container has increased, the pressure exerted by the fluid also has increased. **Figure 19** shows how the force exerted on a brake pedal is transmitted to a fluid.

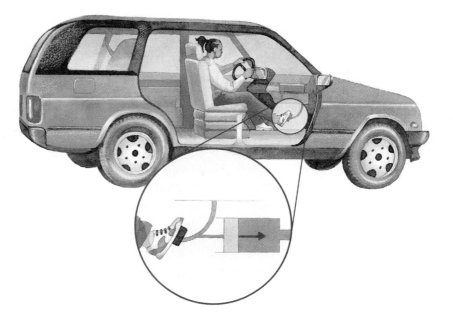

Figure 19
Because the fluid in this piston can't escape, it transmits the force you apply throughout the fluid.

Pascal's Principle

Suppose you fill a plastic bottle with water and screw the cap back on. If you poke a hole in the bottle near the top, water will leak out of the hole. However, if you squeeze the bottle near the bottom, as shown in **Figure 20,** water will shoot out of the hole. When you squeezed the bottle, you applied a force on the fluid. This increased the pressure in the fluid and pushed the water out of the hole faster.

No matter where you poke the hole in the bottle, squeezing the bottle will cause the water to flow faster out of the hole. The force you exert on the fluid by squeezing has been transmitted to every part of the bottle. This is an example of Pascal's principle. According to **Pascal's principle,** when a force is applied to a fluid in a closed container, the pressure in the fluid increases everywhere by the same amount. In other words, the increase in force is transmitted to all parts of the fluid.

Figure 20
When you squeeze the bottle, the pressure you apply is distributed throughout the fluid, forcing the water out the hole.

Hydraulic Systems

Pascal's principle is used in building hydraulic systems like the ones used by car lifts. A **hydraulic system** uses a fluid to increase an applied force. The fluid enclosed in a hydraulic system transfers pressure from one piston to another. An example is shown in **Figure 21.** A force that is applied to the small piston puts pressure on the fluid. The pressure is transmitted throughout the fluid. This causes the force at the bottom of the larger piston to increase.

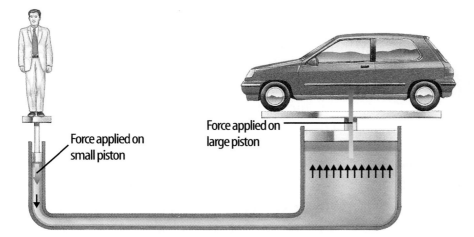

Force applied on small piston

Force applied on large piston

Figure 21
A person standing on the small piston increases the pressure in the fluid. This pressure increase is transmitted to the large piston. The area that is in contact with the fluid in this piston is larger than the area in the smaller piston. Therefore, the force exerted on the large piston is greater than the force applied on the small piston.

Increasing Force What is the force pushing upward on the larger piston? For example, suppose that the area of the small piston is 1 m² and the area of the large piston is 2 m². If you push on the small piston with a force of 10 N, the increase in pressure at the bottom of the small piston is

$$P = F/A$$
$$= (10\ \text{N})/(1\ \text{m}^2)$$
$$= 10\ \text{Pa}$$

According to Pascal's principle, this increase in pressure is transmitted throughout the fluid. This causes the force exerted by the fluid on the larger piston to increase. The increase in the force on the larger piston can be calculated by multiplying both sides of the above formula by *A*.

$$F = P \times A$$
$$= 10\ \text{Pa} \times 2\ \text{m}^2$$
$$= 20\ \text{N}$$

The force pushing upward on the larger piston is twice as large as the force pushing downward on the smaller piston. What happens if the larger piston increases in size? Look at the calculation above. If the area of the larger piston increases to 5 m², the force pushing up on this piston increases to 50 N. So a small force pushing down on the left piston as in **Figure 21** can be made much larger by increasing the size of the piston on the right.

✔ **Reading Check** *How does a hydraulic system increase force?*

Pressure in a Moving Fluid

What happens to the pressure in a fluid if the fluid is moving? Try the following experiment. Place an empty soda can on the desktop and blow to the right of the can, as shown in **Figure 22.** In which direction will the can move?

When you blow to the right of the can, the can moves to the right, toward the moving air. The air pressure exerted on the right side of the can, where the air is moving, is less than the air pressure on the left side of the can, where the air is not moving. As a result, the force exerted by the air pressure on the left side is greater than the force exerted on the right side, and the can is pushed to the right. What would happen if you blew between two empty cans?

Figure 22
By blowing on one side of the can, you decrease the air pressure on that side. Because the pressure on the opposite side is now greater, the can moves toward the side you're blowing on.

Bernoulli's Principle

The reason for the surprising behavior of the can in **Figure 22** was discovered by the Swiss scientist Daniel Bernoulli in the eighteenth century. It is an example of Bernoulli's principle. According to **Bernoulli's principle,** when the speed of a fluid increases, the pressure exerted by the fluid decreases. When you blew across the side of the can, the pressure exerted by the air on that side of the can decreased because the air was moving faster than it was on the other side. As a result, the can was pushed toward the side you blew across.

Chimneys and Bernoulli's Principle In a fireplace the hotter, less dense air above the fire is pushed upward by the cooler, denser air in the room. Wind outside of the house can increase the rate at which the smoke rises. Look at **Figure 23.** Air moving across the top of the chimney causes the air pressure above the chimney to decrease according to Bernoulli's principle. As a result, more smoke is pushed upward by the higher pressure of the air in the room.

Damage from High Winds You might have seen photographs of people preparing for a hurricane by closing shutters over windows or nailing boards across the outside of windows. In a hurricane, the high winds blowing outside the house cause the pressure outside the house to be less than the pressure inside. This difference in pressure can be large enough to cause windows to be pushed out and to shatter.

Hurricanes and other high winds sometimes can blow roofs from houses. When wind blows across the roof of a house, the pressure outside the roof decreases. If the wind outside is blowing fast enough, the outside pressure can become so low that the roof can be pushed off the house by the higher pressure of the still air inside.

Wings and Flight

You might have placed your hand outside the open window of a moving car and felt the push on it from the air streaming past. If you angled your hand so it tilted upward into the moving air, you would have felt your hand pushed upward. If you increased the tilt of your hand, you felt the upward push increase. You might not have realized it, but your hand was behaving like an airplane wing. The force that lifted your hand was provided by a fluid—the air.

Figure 23
The air moving past the chimney lowers the air pressure above the chimney. As a result, smoke is forced up the chimney faster than when air above the chimney is still.

Producing Lift How is the upward force, or lift, on an airplane wing produced? A jet engine pushes the plane forward, or a propeller pulls the plane forward. Air flows over the wings as the plane moves. The wings are tilted upward into the airflow, just like your hand was tilted outside the car window. **Figure 24** shows how the tilt of the wing causes air that has flowed over the wing's upper and lower surfaces to be directed downward.

Lift is created by making the air flow downward. To understand this, remember that air is made of different types of molecules. The motion of these molecules is changed only when a force acts on them. When the air is directed downward, a force is being exerted on the air molecules by the wing.

However, according to Newton's third law of motion, for every action force there is an equal but opposite reaction force. The wing exerts a downward action force on the air. So the air must exert an upward reaction force on the wing. This reaction force is the lift that enables paper airplanes and jet airliners to fly.

Airplane Wings Airplanes have different wing shapes, depending on how the airplane is used. The lift on a wing depends on the amount of air that the wing deflects downward and how fast that air is moving. Lift can be increased by increasing the size or surface area of the wing. A larger wing is able to deflect more air downward.

Look at the planes in **Figure 25.** A plane designed to fly at high speeds, such as a jet fighter, can have small wings. A large cargo plane that carries heavy loads needs large wings to provide a great deal of lift. A glider flies at low speeds and uses long wings that have a large surface area to provide the lift it needs.

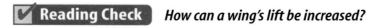 **Reading Check** *How can a wing's lift be increased?*

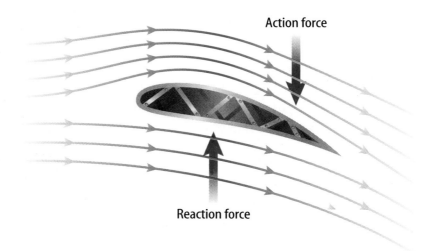

Figure 24
An airplane wing forces air to be directed downward. As a result, the air exerts an upward reaction force on the wing, producing lift.

Action force

Reaction force

Figure 25
Different wing shapes are used for different types of planes. Larger wings provide more lift.

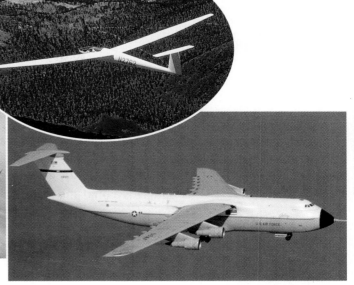

Life Science
INTEGRATION

Birds' Wings A bird's wing provides lift in the same way that an airplane wing does. The wings also act as propellers that pull the bird forward when it flaps its wings up and down. Bird wings also have different shapes depending on the type of flight. Seabirds have long, narrow wings, like the wings of a glider, that help them glide long distances. Forest and field birds, such as pheasants, have short, rounded wings that enable them to take off quickly and make sharp turns. Swallows, swifts, and falcons, which fly at high speeds, have small, narrow, tapered wings like those on a jet fighter.

Section ③ Assessment

1. What is Pascal's principle?

2. What is Bernoulli's principle?

3. Which would you expect to carry more cargo: a plane with small wings or one with large wings? Why?

4. What does a hydraulic lift system use to increase an applied force?

5. **Think Critically** Suppose you are in a small building with a single window during a storm. To keep the roof from being blown off, should you open the window or keep it closed? Explain your answer.

Skill Builder Activities

6. **Drawing Conclusions** When high-speed trains pass each other, windows can be pulled out of their frames. Why? **For more help, refer to the** Science Skill Handbook.

7. **Solving One-Step Equations** A force of 250 N is applied to the small piston of a hydraulic lift, which has an area of 0.01 m². If the large piston has an area of 0.75 m², what force will the fluid exert on it? **For more help, refer to the** Math Skill Handbook.

Activity *Use the Internet*

Barometric Pressure and Weather

The weight of Earth's atmosphere exerts pressure on Earth's surface. The atmosphere is a fluid and flows from one place to another as weather patterns change. Changing weather conditions also cause the atmospheric pressure to change. By analyzing changes in barometric pressure and observing weather conditions, you can make weather predictions.

Recognize the Problem

How do changes in barometric pressure help predict weather patterns?

Form a Hypothesis

What is the current barometric pressure where you are? How would you describe the weather today where you are? What is the weather like in the region to the west of you? To the east of you? What will your weather be like tomorrow? By collecting barometric pressure data and observing weather conditions, you will be able to make a prediction about the next day's weather.

Goals

- **Collect** barometric pressure and other weather data.
- **Compare** barometric pressure to weather conditions.
- **Predict** weather patterns based on barometric pressure, wind speed and direction, and visual conditions.

Data Source

SCIENCE *Online* Go to the Glencoe Science Web site at **science.glencoe.com** to get more information about barometric pressure, weather information, and data collected by other students.

Test Your Hypothesis

Plan

1. Go to **science.glencoe.com** for links to information about weather in the region where you live.

2. Find and record the current barometric pressure and whether the pressure is rising, falling, or remaining steady. Also record the wind speed and direction.

3. **Observe and record** other weather conditions, such as whether rain is falling, the Sun is shining, or the sky is cloudy.

4. Based on the data you collect and your observations, predict what you think tomorrow's weather will be. Record your prediction.

5. Repeat the data collection and observation for a total of five days.

Do

1. Make sure your teacher approves your plan before you start.

2. Go to the Glencoe Science Web site at **science.glencoe.com** to post your data.

Barometric Pressure and Weather Data	
Location of weather station	
Barometric pressure	
Status of barometric pressure	
Wind speed	
Wind direction	
Current weather conditions	
Prediction of tomorrow's weather conditions	

Analyze Your Data

1. Look at your data. What was the weather the day after the barometric pressure increased? The day after the barometric pressure decreased? The day after the barometric pressure was steady?

2. How accurate were your weather predictions?

Draw Conclusions

1. What is the weather to the west of you today? How will that affect the weather in your area tomorrow?

2. What was the weather to the east of you today? How does that compare to the weather in your area yesterday?

3. How does increasing, decreasing, or steady barometric pressure affect the weather?

𝒞ommunicating **Your Data**

Use the data on the Web site to predict the weather where you are two days from now.

"Hurricane"
by John Balaban

Respond to the Reading

1. What kinds of damage did the hurricane cause?

2. Why do you think the poet felt relief when his friend, Elling, arrived?

3. In what part of the country do you think the hurricane occurred?

Near dawn our old live oak sagged over
then crashed on the tool shed
rocketing off rakes paintcans flower pots.

All night, rain slashed the shutters until
it finally quit and day arrived in queer light,
silence, and ozoned air. Then voices calling

as neighbors crept out to see the snapped trees,
leaf mash and lawn chairs driven in heaps
with roof bits, siding, sodden birds, dead snakes.

For days, bulldozers clanked by our houses
in sickening August heat as heavy cranes
scraped the rotting tonnage from the streets.

Then our friend Elling drove in from Sarasota
in his old . . . van packed with candles, with
dog food, cat food, flashlights and batteries . . . .

Understanding Literature

Sense Impressions Images that appeal to the five senses of touch, taste, smell, sight, and sound can be memorable images for a reader. In this poem, John Balaban uses sense impressions to place the reader directly into the poem's environment. For example, the word *crashed* evokes the sense of sound and the term *rotting tonnage* evokes the sense of smell. Do you think the poet is effective in using sense impressions to convey the experience of a hurricane?

Science Connection Because of Bernoulli's principle, when the wind blows above a roof or across the outside of a house, the pressure outside the house or above the roof decreases, but the pressure inside the house doesn't change. In the poem, bits of roof and siding are thrown from the houses. The pressure outside the houses had become lower than the pressure inside the houses because of the high wind speed caused by the hurricane. This caused the siding and roofing to pop off.

Linking Science and Writing

Nature Poem Write a poem describing a natural phenomenon that demonstrates the principles explained in this chapter—forces and fluids. You can describe a natural disaster such as a tornado or any natural occurrence, such as a leaf falling from a tree. Try to use words that evoke at least one of the five sense impressions—touch, taste, smell, sight, and sound.

Career Connection

Naval Architect and Project Engineer

Joan A. Suggs-Cooper is a naval architect and project engineer with the Department of the Navy's Sea Systems Command. Naval architects and project engineers provide the tools needed to design ships and submarines for the U.S. Naval Fleet, Coast Guard, and Military Sealift Command. They think of innovative ways to work together to design the best ships and submarines, and build them faster and cheaper than in the past. Math, science, and English skills are necessary for the designing and communication of their plans. Ms. Suggs-Cooper has a bachelor's degree in civil engineering and a master's degree in technology management from the University of Maryland.

SCIENCE *Online* To learn more about careers in naval science, visit the Glencoe Science Web site at **science.glencoe.com.**

Reviewing Main Ideas

Section 1 Pressure

1. Pressure equals force divided by area. *What advantage does the camel's wide foot have when it is walking on sand?*

2. A fluid is a substance that can flow. Liquids and gases are fluids.

3. Pressure increases with depth and decreases with elevation in a fluid.

4. The pressure exerted by a fluid on a surface is always perpendicular to the surface.

Section 2 Why do objects float?

1. A buoyant force is an upward force exerted on all objects placed in a fluid.

2. The buoyant force depends on the shape of the object.

3. An object placed in a fluid displaces some of the fluid. According to Archimedes' principle, the buoyant force on the object is equal to the weight of the fluid displaced by the object.

4. An object floats when the buoyant force exerted by the fluid is equal to the object's weight.

5. An object will float if it is less dense than the fluid it is placed in. *Why is the ball floating in the photo below?*

Section 3 Doing Work with Fluids

1. Pascal's principle states that the pressure applied at any point to a confined fluid is transmitted unchanged throughout the fluid. Hydraulic systems are applications of Pascal's principle.

2. Bernoulli's principle states that when the velocity of a fluid increases, the pressure exerted by the fluid decreases.

3. A wing provides lift by forcing air downward. *How does the size of a wing affect the lift?*

FOLDABLES
Reading & Study Skills

After You Read

Using the information on your Foldable, write about the effect of force on liquids and gases under the left and right tabs.

Visualizing Main Ideas

Fill in the table below.

Relationships Among Forces and Fluids		
Idea	**What does it relate?**	**How?**
Density	mass and volume	
Pressure		force/area
Archimedes' principle	buoyant force and weight of fluid that is displaced	
Bernoulli's principle		velocity increases, pressure decreases
Pascal's principle	pressure applied to enclosed fluid at one point and pressure at other points in a fluid	

Vocabulary Review

Vocabulary Words

a. Archimedes' principle
b. Bernoulli's principle
c. buoyant force
d. density
e. fluid
f. hydraulic system
g. Pascal's principle
h. pressure

THE PRINCETON REVIEW Study Tip

Listening is a learning tool, too. Try recording a reading of your notes on a tape and replaying it for yourself a few times a week.

Using Vocabulary

Answer each of the following questions using complete sentences.

1. How would you describe a substance that can flow?

2. When the area over which a force is applied decreases, what increases?

3. What principle relates the weight of displaced fluid to the buoyant force?

4. How is a fluid used to lift heavy objects?

5. If you increase an object's mass but not its volume, what have you changed?

6. How is a log able to float in a river?

7. What principle explains why hurricanes can blow the roof off a house?

Chapter 3 Assessment

Checking Concepts

Choose the word or phrase that best answers the question.

1. What happens to pressure when the force that is exerted on a surface decreases?
   A) decreases
   B) increases
   C) stays the same
   D) any of the above

2. If the pressure that is exerted by a moving fluid decreases, its speed does what?
   A) decreases
   B) increases
   C) stays the same
   D) any of the above

3. What is the pressure due to a force of 100 N on an area 4 m^2?
   A) 4 Pa
   B) 20 Pa
   C) 25 Pa
   D) 400 Pa

4. If force increased and pressure remained the same, what must have happened?
   A) density decreased
   B) density increased
   C) area decreased
   D) area increased

5. A piece of cork has a density of 0.23 g/cm^3 and a mass of 46 g. What is its volume?
   A) 0.005 cm^3
   B) 10.6 cm^3
   C) 100 cm^3
   D) 200 cm^3

6. As a rock sinks deeper into a pond, the buoyant force on it does what?
   A) increases
   B) stays the same
   C) decreases
   D) disappears

7. As you drive down from a high mountain, what happens to air pressure?
   A) It decreases.
   B) It increases.
   C) It increases, then decreases.
   D) It stays the same.

8. The weight of the fluid displaced by an object is equal to which of the following?
   A) the weight of the object
   B) the density of the object
   C) the buoyant force on the object
   D) the volume of the object

9. The buoyant force on a floating object is equal to which of the following?
   A) weight of object
   B) volume of object
   C) density of fluid
   D) density of object

10. What would you expect to happen if you blew hard between two empty soda cans?
    A) They will move together.
    B) They will move apart.
    C) They will move toward you.
    D) They won't move.

Thinking Critically

11. A sandbag is dropped from a hot-air balloon and the balloon rises. Explain, using the buoyant force of air.

12. Explain whether or not this statement is correct: Heavy objects sink, and light objects float.

13. A rock is attached to a spring scale that reads 10 N. If the rock is submerged in water, the scale reads 6 N. What is the buoyant force on the submerged rock?

14. Explain why a leaking boat sinks.

15. Explain why the direction of the buoyant force on a submerged cube is upward and not left or right.

Developing Skills

16. **Solving One-Step Equations** A hydraulic lift with a large piston area of 4 m^2 exerts a force of 50,000 N. If the smaller piston has an area of 0.05 m^2, what is the force on it?

17. **Recognizing Cause and Effect** A steel tank and a balloon are the same size and contain the same amount of helium. Explain why the balloon rises and the steel tank doesn't.

18. **Classifying** Classify which of the following substances will and will not float in water.

Material Density	
Substance	**Density (g/cm³)**
Ice	0.92
Lead	11.34
Balsa wood	0.12
Sugar	1.59

19. **Making and Using Graphs** Graph the pressure exerted by a 75-kg person wearing different shoes with areas of 0.01 m², 0.02 m², 0.03 m², 0.04 m², and 0.05 m². Plot pressure on the vertical axis and area on the horizontal axis.

Performance Assessment

20. **Experiment** Partially fill a plastic dropper with water until it is heavy enough to float just below the water's surface. With the dropper inside, completely fill a plastic bottle with water and seal the top. When you squeeze the bottle, the dropper will sink. Confirm this. What happens to water pressure in the bottle? In the dropper? To the air in the dropper? Use all of these to explain what you observe.

21. **Oral Presentation** Research the different wing designs in animals or aircraft. Present your results to the class.

TECHNOLOGY

Go to the Glencoe Science Web site at **science.glencoe.com** or use the **Glencoe Science CD-ROM** for additional chapter assessment.

THE PRINCETON REVIEW **Test Practice**

Jonah completed the following experiment in his home in preparation for a science project.

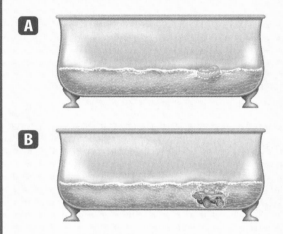

A

B

Study the above illustrations and answer the following questions.

1. Based on the diagram, why is the toy truck on the bottom of tub B?
   A) The density of the truck is less than the density of the water.
   B) The density of the truck is greater than the density of the water.
   C) The density of the truck equals the density of the water.
   D) The weight of the truck is equal to the weight of the water in the tub.

2. What would happen if Jonah removed half of the water from tub A and added it to tub B?
   F) The truck would float, but the soap would sink.
   G) The truck would sink, and the soap also would sink.
   H) The truck and the soap would float.
   J) The truck would sink, and the soap would float.

Work and Simple Machines

Machines enable people to accomplish many different tasks, from eating a meal to building a skyscraper. In this picture, machines such as cranes and trucks are being used to help construct buildings. In this chapter, you will learn about work and what simple machines are. You also will learn how simple machines make doing work easier.

What do you think?

Science Journal Look at the picture below with a classmate. Discuss what you think this might be or what is happening. Here's a hint: *The other six legs are just as hairy.* Write down your answer or best guess in your Science Journal.

EXPLORE ACTIVITY

Two of the world's greatest structures were built using different tools. The Great Pyramid at Giza in Egypt was built nearly 5,000 years ago using blocks of limestone moved into place by hand with ramps and levers. In comparison, the Sears Tower in Chicago was built in 1973 using tons of steel that were hoisted into place by gasoline-powered cranes. How do machines such as ramps, levers, and cranes change the forces needed to do a job?

Compare forces

1. Place a ruler on an eraser. Place a book on one end of the ruler.

2. Using one finger, push down on the free end of the ruler to lift the book.

3. Repeat the experiment, placing the eraser in various positions beneath the ruler. Observe how much force is needed in each instance to lift the book.

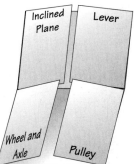

Observe

In your Science Journal, describe your observations. How did changing the distance between the book and the eraser affect the force needed to lift the book?

Before You Read

FOLDABLES
Reading & Study Skills

Making a Main Ideas Study Fold Make the following Foldable to help you identify the main ideas or major topics of work and simple machines.

1. Place a sheet of paper in front of you so the long side is at the top. Fold the paper in half from the left side to the right side and then unfold.

2. Fold each side in to the centerfold line to divide the paper into fourths. Fold the paper in half from top to bottom and unfold.

3. Through the top thickness of paper, cut along both of the middle fold lines to form four tabs as shown. Label each tab *Inclined Plane, Wheel and Axle, Lever,* and *Pulley* as shown.

4. Before you read the chapter, fold your Foldable in half to make a book. Title the book *Work* and define work under the title. As you read the chapter, write information under the tabs about the four tools on your Foldable.

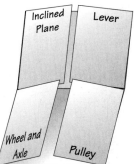

Work and Power

As You Read

What **You'll Learn**

■ **Recognize** when work is done.
■ **Calculate** how much work is done.
■ **Explain** the relation between work and power.

Vocabulary
work
power

Why **It's Important**
If you understand work, you can make your work easier.

What is work?

What does the term *work* mean to you? You might think of household chores, a job at an office, a factory, a farm, or the homework you do after school. In science, the definition of work is more specific. **Work** is done when a force causes an object to move in the same direction that the force is applied.

Can you think of a way in which you did work today? Maybe it would help to know that you do work when you lift your books, turn a doorknob, raise window blinds, or write with a pen or pencil. You also do work when you walk up a flight of stairs or open and close your school locker. In what other ways do you do work every day?

Work and Motion Your teacher has asked you to move a box of books to the back of the classroom. Try as you might, though, you just can't budge the box because it is too heavy. Although you exerted a force on the box and you feel tired from it, you have not done any work. In order for you to do work, two things must occur. First, you must apply a force to an object. Second, the object must move in the same direction as your applied force. You do work on an object only when the object moves as a result of the force you exert. The girl in **Figure 1** might think she is working by holding the bags of groceries. However, if she is not moving, she is not doing any work because she is not causing something to move.

☑ **Reading Check** *Why don't you do work when you hold a baby?*

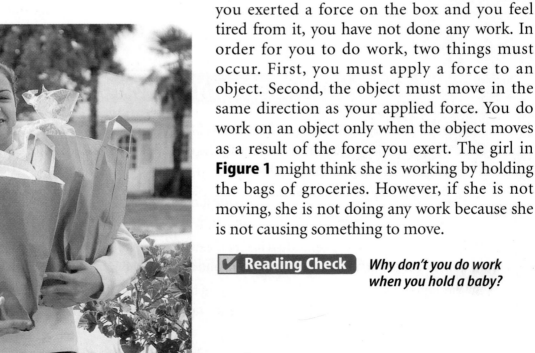

Figure 1
This girl is holding bags of groceries, yet she isn't doing any work. *Why?*

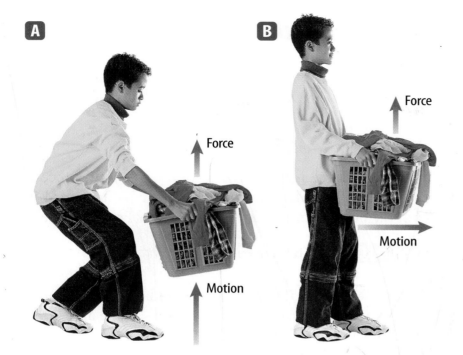

A **B**

Force

Force

Motion

Motion

Figure 2
To do work, an object must move
in the direction a force is applied.
A The boy's arms do work when
they exert an upward force on
the basket and the basket moves
upward. **B** The boy's arms still
exert an upward force on the
basket. But when the boy walks
forward, no work is done by his
arms.

Applying Force and Doing Work Picture yourself lifting
the basket of clothes in **Figure 2A.** You can feel your arms exerting a force upward as you lift the basket, and the basket moves
upward in the direction of the force your arms applied. Therefore, your arms have done work. Now suppose you carry the
basket forward, as in **Figure 2B.** You can still feel your arms
applying an upward force on the box to keep it from falling, but
now the box is moving forward instead of upward. Because the
direction of motion is not in the same direction of the force
applied by your arms, no work is done by your arms.

Force in Two Directions Sometimes only part of the force
you exert moves an object. Think about what happens when you
push a lawn mower. You push at an angle to the ground as shown
in **Figure 3.** Part of the force is to the right and part of the force
is downward. Only the part of the force that is in the same direction as the motion of the mower—to the right—does work.

Forward force

Total
force

Downward
force

Motion

Figure 3
When you exert a force at an
angle, only part of your force
does work—the part that is
in the same direction as the
motion of the object.

Calculating Work

Work is done when a force makes an object move. More work is done when the force is increased or the object is moved a greater distance. The work done can be calculated from this equation:

$$\text{Work} = \text{force} \times \text{distance}$$

In SI units, force is measured in newtons and distance is measured in meters. The unit for work is the joule, named in honor of the nineteenth-century scientist James Prescott Joule.

Reading Check *What is the SI unit for work?*

Work and Distance Suppose you give a book a push and it slides across a table. To calculate the work you did, the distance in the above equation is not the distance the book moved. The distance in the work equation is the distance an object moves while the force is being applied. So the distance in the work equation is the distance the book moved while you were pushing.

Math Skills Activity

Calculating Work

Example Problem

A painter lifts a can of paint that weighs 40 N a distance of 2 m. How much work does she do? Hint: to lift a can weighing 40 N, the painter must exert a force of 40 N.

Solution

1 *This is what you know:*

force = 40 N
distance = 2 m

2 *This is what you need to know:*   work

3 *This is the equation you need to use:*   work = force × distance

4 *Substitute the known values into the equation:*   work = 40 N × 2 m = 80 J

Check your answer by dividing it by the distance. Did you calculate the same force that was given?

> **Practice Problem**
>
> As you push a lawn mower, the horizontal force is 300 N. If you push the mower a distance of 500 m, how much work do you do?

For more help, refer to the Math Skill Handbook.

What is power?

What does it mean to be powerful? To understand power, imagine two weightlifters lifting the same weight and suppose they lift the weight the same distance. Because they exert the same upward force and move the weight the same distance, each does the same amount of work.

Suppose one weightlifter lifted the weight in 3 s, while the other struggled to lift it in 10 s. You might say that the first weightlifter is stronger, because she lifted the weight in less time. You could also say that she is more powerful. In science, **power** is how quickly work is done. Something has more power or is more powerful if it can do more work in a certain amount of time.

Calculating Power Power can be calculated by dividing the amount of work done by the time needed to do the work. Power can be calculated by this formula:

$$\text{Power} = \frac{\text{work done}}{\text{time needed}}$$

In SI units, the unit of power is the watt, in honor of James Watt, a nineteenth-century British scientist who invented a practical version of the steam engine.

SCIENCE *Online*

Research Visit the Glencoe Science Web site at **science.glencoe.com** for more information about James Watt and his steam engine. Find out why the unit of power was named after this inventor. Summarize your findings in a brief report.

Math Skills Activity

Calculating Power

Example Problem
You do 200 J of work in 12 s. How much power did you use?

Solution

1 *This is what you know:*   work done = 200 J
     time needed = 12 s

2 *This is what you need to find:*   power

3 *This is the equation you need to use:*   power = work done/time needed

4 *Substitute the known values into the equation:*   power = 200 J/12 s = 17 watt

Check your answer by dividing it by the work done. Did you calculate the same time that was given?

Practice Problem

In the course of a short race, a car does 500,000 J of work in 7 s. What is the power of the car during the race?

For more help, refer to the Math Skill Handbook.

Measuring Work and Power

Procedure

1. Weigh yourself on a **scale**.
2. Multiply your weight in pounds by 4.45 to convert your weight to newtons.
3. Measure the vertical height of a **ramp or stairway**. **WARNING:** *Make sure the ramp or stairway is clear of all objects.*
4. Time yourself walking slowly and quickly up the ramp or stairway.

Analysis

1. Calculate and compare the work done and power used in each case.
2. How would the work done and power used change if your weight were twice as large?

Work and Energy If you push a chair and make it move, you do work on the chair. You also change the energy of the chair. Recall that when something is moving it has energy of motion, or kinetic energy. By making the chair move, you increase its kinetic energy.

You also change the energy of an object when you do work and lift it higher. An object has potential energy that increases when it is higher above Earth's surface. By lifting an object, you do work and increase its potential energy.

Power and Energy When you do work on an object you increase the energy of the object. Where does this energy come from? Because energy can never be created or destroyed, if the object gains energy then you must lose energy. When you do work on an object you transfer energy to the object, and your energy decreases. The amount of work done is the amount of energy transferred. So power is also equal to the amount of energy transferred in a certain amount of time.

$$\text{Power} = \frac{\text{energy transferred}}{\text{time needed}}$$

Sometimes energy can be transferred even when no work is done, such as when heat flows from a warm to a cold object. In fact, there are many ways energy can be transferred even if no work is done. Power is always the rate at which energy is transferred, or the amount of energy transferred divided by the time needed.

Section ① Assessment

1. What conditions must be met for work to be done?
2. How much work was done to lift a 1,000-kg block to the top of the Great Pyramid, 146 m above the ground?
3. How is power related to work?
4. How much power, in watts, is needed to cut a lawn in 50 min if the work involved is 100,000 J?
5. **Think Critically** Suppose you are pulling a wagon at an angle. How can you make your task easier?

Skill Builder Activities

6. **Comparing and Contrasting** Which example involves more power: 200 J of work done in 20 s or 50 J of work done in 4 s? Explain your answer. **For more help, refer to the** Science Skill Handbook.

7. **Solving One-Step Equations** A 7,460-W engine is used to lift a beam weighing 9,800 N up 145 m. How much work must the motor do to lift this beam at constant speed? How much more work must be done to lift it 290 m? **For more help, refer to the** Math Skill Handbook.

Activity

Building the Pyramids

Imagine moving 2.3 million blocks of limestone, each weighing more than 1,000 kg. That is exactly what the builders of the Great Pyramid at Giza did. Although no one knows for sure exactly how they did it, they probably pulled the blocks most of the way. What could they have done to make their work easier?

What You'll Investigate
How is the force needed to lift a block related to the distance it travels?

Materials
wood block   thin notebooks
tape   meterstick
spring scale   several books
ruler

Goals
- **Compare** the force needed to lift a block with the force needed to pull it up a ramp.

Safety Precautions

Procedure

1. Stack several books together on a tabletop to model a half-completed pyramid. Measure the height of the books in centimeters. Record the height on the first row of the data table under *Distance.*

2. Use the wood block as a model for a block of stone. Use tape to attach the block to the spring scale.

3. Place the block on the table and lift it straight up the side of the stack of books until the top of the block is even with the top of the books. Record the force shown on the scale in the data table under *Force.*

Work Done Using Different Ramps		
Distance (cm)	Force (N)	Work (J)

4. **Arrange** a notebook so that one end is on the stack of books and the other end is on the table. Measure the length of the notebook and record this length as distance in the second row of the data table under *Distance.*

5. **Measure** the force needed to pull the block up the ramp. Record the force in the data table.

6. Repeat steps 4 and 5 using a longer notebook to make the ramp longer.

7. **Calculate** the work done in each row of the data table.

Conclude and Apply

1. How much work did you do in each case?

2. What happened to the force needed as the length of the ramp increased?

3. How could the builders of the pyramids have designed their task to use less force than they would lifting the blocks straight up? Draw a diagram to support your answer.

*C*ommunicating Your Data

Add your data to that found by other groups. **For more help, refer to the** Science Skill Handbook.

② Using Machines

As You Read

What You'll Learn
- **Explain** how a machine makes work easier.
- **Calculate** the mechanical advantages and efficiency of a machine.
- **Explain** how friction reduces efficiency.

Vocabulary
input force
output force
mechanical advantage
efficiency

Why It's Important
Machines can't change the amount of work you need to do, but they can make doing work easier.

What is a machine?

Did you use a machine today? When you think of a machine you might think of a device, such as a car, with many moving parts powered by an engine or an electric motor. But if you used a pair of scissors or a broom, or cut your food with a knife, you used a machine. A machine is simply a device that makes doing work easier. Even a sloping surface can be a machine.

Mechanical Advantage

Even though machines make work easier, they don't decrease the amount of work you need to do. Instead, a machine changes the way in which you do work. When you use a machine, you exert a force over some distance. For example, you exert a force to move a rake or lift the handles of a wheelbarrow. This force is called the effort force, or the **input force.** The work you do on the machine is equal to the input force times the distance over which your force moves the machine. The work that you do on the machine is the input work.

The machine also does work by exerting a force to move an object over some distance. A rake, for example, exerts a force to move leaves. Sometimes this force is called the resistance force because the machine is trying to overcome some resistance. This force also can be called the **output force.** The work that the machine does is the output work. **Figure 4** shows how a machine transforms input work to output work.

When you use a machine, the output work can never be greater than the input work. So what is the advantage of using a machine? A machine makes work easier by changing the amount of force you need to exert, the distance over which the force is exerted, or the direction in which you exert your force.

Figure 4
No matter what type of machine is used, the output work is never greater than the input work.

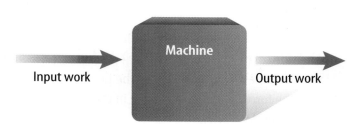

Input work · Machine · Output work

Changing Force Work is equal to force times distance. If work stays the same, what happens to force if you exert a force over a longer distance? You can exert a smaller force. Some machines make work easier by allowing you to exert a smaller force over a longer distance.

The mechanical advantage of a machine compares the input force to the output force. **Mechanical advantage** is the number of times the input force is multiplied by a machine.

$$\text{Mechanical advantage} = \frac{\text{output force}}{\text{input force}}$$

For example, suppose that using a pulley system takes you only 300 N to lift a piano that weighs 1,500 N. To lift the piano, the pulley system exerts an upward force of 1,500 N to overcome the downward pull of gravity. This is the output force. The force you exert on the pulley system in the input force, which is 300 N. So the mechanical advantage of the pulley system is five.

SCIENCE *Online*

Research Visit the Glencoe Science Web site at **science.glencoe.com** for more information about early types of tools and how they took advantage of simple machines. Design a poster relating several ancient tools to the simple machines you have studied.

✔ **Reading Check** *What is the mechanical advantage of a machine?*

Math Skills Activity

Calculating Mechanical Advantage

Example Problem

To pry the lid off a paint can, you apply a force of 50 N to the handle of a screwdriver. What is the mechanical advantage of the screwdriver if it applies a force of 500 N to the lid?

Solution

1 *This is what you know:*   output force = 500 N
                                    input force = 50 N

2 *This is what you need to find:*   mechanical advantage

3 *This is the equation you need to use:*   mechanical advantage = output force/input force

4 *Substitute the known values:*   mechanical advantage = (500 N) / (50 N) = 10.

Check your answer by multiplying it by the input force. Do you calculate the same output force that was given?

Practice Problem

To open a bottle, you apply a force of 50 N to a bottle opener. The bottle opener applies a force of 775 N to the bottle cap. What is the mechanical advantage of the bottle opener?

For more help, refer to the Math Skill Handbook.

Figure 5

A When you rake leaves, you move your hands a short distance but the end of the rake moves over a longer distance. **B** Sometimes it is easier to exert your force in a certain direction. This boy would rather pull down on the rope to lift the flag than to climb to the top of the pole and pull up.

Changing Distance Some machines allow you to exert your force over a shorter distance. In these machines, the output force is less than the input force. The rake in **Figure 5A** is this type of machine. You move your hands a small distance at the top of the handle, but the bottom of the rake moves a greater distance as it moves the leaves. The mechanical advantage of this type of machine is less than one because the output force is less than the input force.

Changing Direction Sometimes it is easier to apply a force in a certain direction. For example, it is easier to pull down on the rope in **Figure 5B** than to pull up on it. Some machines enable you to change the direction of the input force. In these machines neither the force nor the distance is changed. The mechanical advantage of this type of machine is equal to one because the output force is equal to the input force. The three ways machines make doing work easier are summarized in **Figure 6**.

Figure 6

Machines are useful because they can **A** increase force, **B** increase distance, or **C** change the direction in which a force is applied.

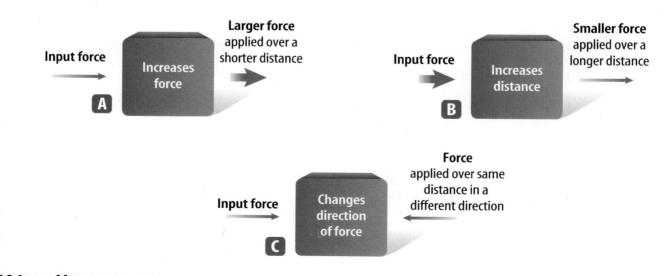

Input force → **A** Increases force → Larger force applied over a shorter distance

Input force → **B** Increases distance → Smaller force applied over a longer distance

Input force → **C** Changes direction of force ← Force applied over same distance in a different direction

Efficiency

A machine doesn't increase the input work. For a real machine, the output work done by the machine is always less than the input work that you do on the machine. Remember that anytime two surfaces slide past each other, friction resists their motion. In real machines some input work is always used to overcome friction, so the parts of the machine can move.

The ability of a machine to convert the input work to output work is called the machine's **efficiency.** Efficiency is described as a percent.

$$\text{Efficiency} = \frac{\text{output work}}{\text{input work}} \times 100\%$$

An ideal machine has an efficiency of 100 percent. The efficiency of a real machine is always less than 100 percent, because some work is converted into heat by friction. When friction is reduced, the efficiency of a machine increases.

> ✔ **Reading Check** *Why is the efficiency of a real machine less than 100 percent?*

Math Skills Activity

Calculating Efficiency

Example Problem

Using a pulley system, a crew does 7,500 J of work to load a box that requires 4,500 J of work. What is the efficiency of the pulley system?

Solution

1 *This is what you know:*   work output = 4,500 J
                                   work input = 7,500 J

2 *This is what you need to find:*   efficiency

3 *This is the equation you need to use:*   efficiency = output work/input work × 100%

4 *Substitute the known values:*   efficiency = (4,500 J)/(7,500 J) × 100% = 60%

Check your answer by multiplying it by the work input.
Do you calculate the same work output that was given?

> **Practice Problem**
>
> You do 100 J of work in pulling out a nail with a claw hammer. If the hammer does 70 J of work, what is the hammer's efficiency?

For more help, refer to the Math Skill Handbook.

Figure 7
Lubrication can reduce the friction between two surfaces.

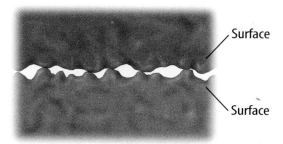

A Two surfaces in contact can stick together where the high spots on each surface come in contact.

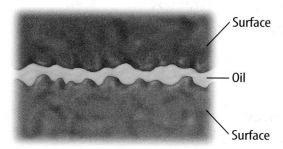

B Adding oil or another lubricant separates the surface so that fewer high spots make contact.

Friction To help understand friction, imagine pushing a heavy box up a ramp. As the box begins to move, the bottom surface of the box slides across the top surface of the ramp. Neither surface is perfectly smooth—each has high spots and low spots, as shown in **Figure 7.**

As the two surfaces slide past each other, high spots on the two surfaces come in contact. At these contact points, shown in **Figure 7A,** atoms and molecules can bond together. This makes the contact points stick together. The attractive forces between all the bonds in the contact points added together is the frictional force that tries to keep the two surfaces from sliding past each other.

To keep the box moving, a force must be applied to break the bonds between the contact points. Even after these bonds are broken and the box moves, new bonds form as different parts of the two surfaces come into contact. So as you keep pushing the box, part of the force that you exert is used to break the bonds that keep forming between the contact points.

Friction and Efficiency One way to reduce friction between two surfaces is to add oil. **Figure 7B** shows how oil fills the gaps between the surfaces, and keeps many of the high spots from making contact. Because there are fewer contact points between the surfaces, the force of friction is reduced. More of the input work then is converted to output work by the machine.

Section 2 Assessment

1. What are three ways in which machines make work easier?
2. How can you find the mechanical advantage of a machine?
3. You do 150 J of work on a machine and the machine does 90 J of work as a result. What is the efficiency of the machine?
4. Explain how friction reduces the efficiency of machines.
5. **Think Critically** Can a machine be useful even if its mechanical advantage is less than one? Explain and give an example.

Skill Builder Activities

6. **Comparing and Contrasting** How does the efficiency of an ideal machine compare with that of a real machine? **For more help, refer to the** Science Skill Handbook.

7. **Using an Electronic Spreadsheet** On a computer, create a spreadsheet that calculates work from force and distance. Input a value for work and then input several different force values. How does the distance change if the work stays the same and the force decreases? **For more help, refer to the** Technology Skill Handbook.

Simple Machines

What is a simple machine?

What do you think of when you hear the word *machine?* Many people think of machines as complicated devices such as cars, elevators, or computers. However, some machines are as simple as a hammer, shovel, or ramp. A **simple machine** is a machine that does work with only one movement. The six simple machines are the inclined plane, lever, wheel and axle, screw, wedge, and pulley. A machine made up of a combination of simple machines is called a **compound machine.** A can opener is a compound machine. The bicycle in **Figure 8** is a familiar example of another compound machine.

Inclined Plane

Ramps might have enabled the ancient Egyptians to build their pyramids. To move limestone blocks weighing more than 1,000 kg each, archaeologists hypothesize that the Egyptians built enormous ramps. A ramp is a simple machine known as an inclined plane. An **inclined plane** is a flat, sloped surface. Less force is needed to move an object from one height to another using an inclined plane than is needed to lift the object. As the inclined plane becomes longer, the force needed to move the object becomes smaller.

***What* You'll Learn**

- **Distinguish** among the different simple machines.
- **Describe** how to find the mechanical advantage of each simple machine.

Vocabulary

simple machine	screw
compound machine	lever
inclined plane	wheel and axle
wedge	pulley

***Why* It's Important**

Simple machines make up all machines.

Figure 8
Devices that use combinations of simple machines, such as this bicycle, are called compound machines.

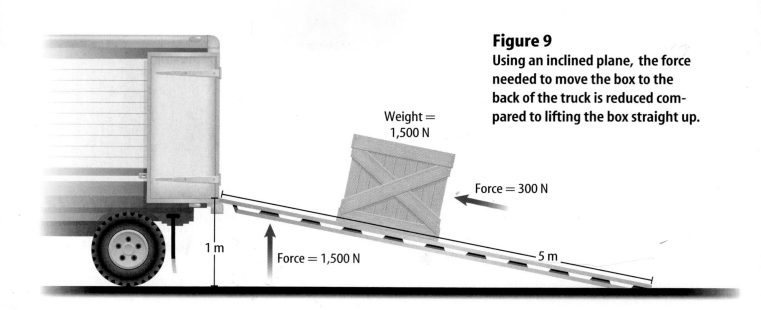

Figure 9
Using an inclined plane, the force needed to move the box to the back of the truck is reduced compared to lifting the box straight up.

Weight = 1,500 N

Force = 300 N

1 m

Force = 1,500 N

5 m

Using Inclined Planes Imagine having to lift a box weighing 1,500 N to the back of a truck that is 1 m off the ground. You would have to exert a force of 1,500 N, the weight of the box, over a distance of 1 m, which equals 1,500 J of work. Now suppose that instead you use a 5-m-long ramp, as shown in **Figure 9.** The amount of work you need to do does not change. You still need to do 1,500 J of work. However, the distance over which you exert your force becomes 5 m. You can calculate the force you need to exert by dividing both sides of the equation for work by distance.

$$\text{Force} = \frac{\text{work}}{\text{distance}}$$

If you do 1,500 J of work by exerting a force over 5 m, the force is only 300 N. Because you exert the input force over a distance that is five times as long, you can exert a force that is five times less.

The mechanical advantage of an inclined plane is the length of the inclined plane divided by its height. In this example, the ramp has a mechanical advantage of 5.

Wedge An inclined plane that moves is called a **wedge.** A wedge can have one or two sloping sides. The knife shown in **Figure 10** is an example of a wedge. An axe and certain types of doorstop are also wedges. Just as for an inclined plane, the mechanical advantage of a wedge increases as it becomes longer and thinner.

Figure 10
This chef's knife is a wedge that slices through food.

Figure 11
Wedge-shaped teeth help tear food.

A Your front teeth help tear an apple apart.

B The wedge-shaped teeth of this Tyrannosaurus Rex show that it was a carnivore.

Life Science INTEGRATION

Wedges in Your Body You have wedges in your body. The bite marks on the apple in **Figure 11A** show how your front teeth are wedge shaped. A wedge changes the direction of the applied effort force. As your push your front teeth into the apple, the downward effort force is changed by your teeth into a sideways force that pushes the skin of the apple apart.

The teeth of meat eaters, or carnivores, are more wedge shaped than the teeth of plant eaters, or herbivores. The teeth of carnivores are used to cut and rip meat, while herbivores' teeth are used for grinding plant material. By examining the teeth of ancient animals, such as the dinosaur in **Figure 11B,** scientists can determine what the animal ate when it was living.

The Screw Another form of the inclined plane is a screw. A **screw** is an inclined plane wrapped around a cylinder or post. The inclined plane on a screw forms the screw threads. Just like a wedge changes the direction of the effort force applied to it, a screw also changes the direction of the applied force. When you turn a screw, the force applied is changed by the threads to a force that pulls the screw into the material. Friction between the threads and the material holds the screw tightly in place. The mechanical advantage of the screw is the length of the inclined plane wrapped around the screw divided by the length of the screw. The more tightly wrapped the threads are, the easier it is to turn the screw. Examples of screws are shown in **Figure 12.**

Figure 12
The thread around a screw is an inclined plane. Many familiar devices use screws to make work easier.

✓ **Reading Check** *How are screws related to the inclined plane?*

Figure 13

The mechanical advantage of a lever changes as the position of the fulcrum changes. The mechanical advantage increases as the fulcrum is moved closer to the output force.

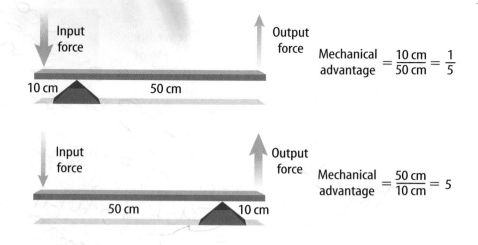

Input force   Output force

10 cm   50 cm

$$\text{Mechanical advantage} = \frac{10\text{ cm}}{50\text{ cm}} = \frac{1}{5}$$

Input force   Output force

50 cm   10 cm

$$\text{Mechanical advantage} = \frac{50\text{ cm}}{10\text{ cm}} = 5$$

Figure 14

A faucet handle is a wheel and axle. A wheel and axle is similar to a circular lever. The center is the fulcrum, and the wheel and axle turn around it. *How can you increase the mechanical advantage of a wheel and axle?*

Wheel

Axle

Input force

Output force

Lever

You step up to the plate. The pitcher throws the ball and you swing your lever to hit the ball? That's right! A baseball bat is a type of simple machine called a lever. A **lever** is any rigid rod or plank that pivots, or rotates, about a point. The point about which the lever pivots is called a fulcrum.

The mechanical advantage of a lever is found by dividing the distance from the fulcrum to the input force by the distance from the fulcrum to the output force, as shown in **Figure 13.** When the fulcrum is closer to the output force than the input force, the mechanical advantage is greater than one.

Levers are divided into three classes according to the position of the fulcrum with respect to the input force and output force. **Figure 15** shows examples of three classes of levers.

Wheel and Axle

Do you think you could turn a doorknob easily if it were a narrow rod the size of a pencil? It might be possible, but it would be difficult. A doorknob makes it easier for you to open a door because it is a simple machine called a wheel and axle. A **wheel and axle** consists of two circular objects of different sizes that are attached in such a way that they rotate together. As you can see in **Figure 14,** the larger object is the wheel and the smaller object is the axle.

The mechanical advantage of a wheel and axle is usually greater than one. It is found by dividing the radius of the wheel by the radius of the axle. For example, if the radius of the wheel is 12 cm and the radius of the axle is 4 cm, the mechanical advantage is 3.

Figure 15

Levers are among the simplest of machines, and you probably use them often in everyday life without even realizing it. A lever is a bar that pivots around a fixed point called a fulcrum. As shown here, there are three types of levers—first class, second class, and third class. They differ in where two forces—an input force and an output force—are located in relation to the fulcrum.

▲ Fulcrum

▼ Input force

▲ Output force

In a first-class lever, the fulcrum is between the input force and the output force. First-class levers, such as scissors and pliers, multiply force or distance depending on where the fulcrum is placed. They always change the direction of the input force, too.

First-class lever

In a second-class lever, such as a wheelbarrow, the output force is between the input force and the fulcrum. Second-class levers always multiply the input force but don't change its direction.

Second-class lever

Third-class lever

In a third-class lever, such as a baseball bat, the input force is between the output force and the fulcrum. For a third-class lever, the output force is less than the input force, but is in the same direction.

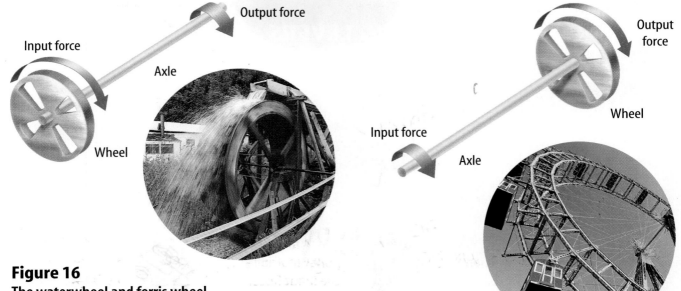

Figure 16
The waterwheel and ferris wheel are examples of devices that rely on a wheel and axle. *How are they alike and how are they different?*

Using Wheels and Axles In some devices, the input force is used to turn the wheel and the output force is exerted by the axle. Because the wheel is larger than the axle, the mechanical advantage is greater than one. So the output force is greater than the input force. A doorknob, a steering wheel, and a screwdriver are examples of this type of wheel and axle.

In other devices, the input force is applied to turn the axle and the output force is exerted by the wheel. Then the mechanical advantage is less than one and the output force is less than the input force. A fan and a ferris wheel are examples of this type of wheel and axle. **Figure 16** shows an example of each type of wheel and axle.

Pulley

To raise a sail, a sailor pulls down on a rope. The rope uses a simple machine called a pulley to change the direction of the force needed. A **pulley** consists of a grooved wheel with a rope or chain wrapped around it.

Fixed Pulleys Some pulleys, such as the one on a sail, a window blind, or a flagpole, are attached to a structure above your head. When you pull down on the rope, you pull something up. This type of pulley, called a fixed pulley, does not change the force you exert or the distance over which you exert it. Instead, it changes the direction in which you exert your force, as shown in **Figure 17A**. The mechanical advantage of a fixed pulley is 1.

✔ **Reading Check** *How does a fixed pulley affect the input force?*

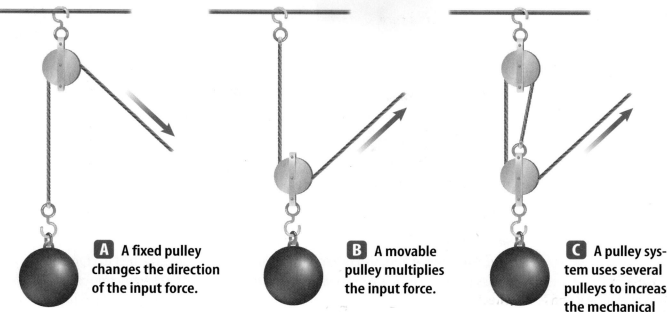

A A fixed pulley changes the direction of the input force.

B A movable pulley multiplies the input force.

C A pulley system uses several pulleys to increase the mechanical advantage.

Movable Pulleys Another way to use a pulley is to attach it to the object you are lifting, as shown in **Figure 17B.** This type of pulley, called a movable pulley, allows you to exert a smaller force to lift the object. The mechanical advantage of a movable pulley is always 2.

More often you will see combinations of fixed and movable pulleys. Such a combination is called a pulley system. The mechanical advantage of a pulley system is equal to the number of sections of rope pulling up on the object. For the pulley system shown in **Figure 17C** the mechanical advantage is 3.

Figure 17
Pulleys can change force and direction.

Section 3 Assessment

1. Define simple and compound machines in your own words.

2. Describe four different simple machines.

3. Why can a machine with a mechanical advantage less than one be useful?

4. How does the mechanical advantage of a wheel and axle change as the size of the wheel increases?

5. **Think Critically** The Great Pyramid is 146 m high. How long would a ramp from the top of the pyramid to the ground need to be to have a mechanical advantage of 4?

Skill Builder Activities

6. **Comparing and Contrasting** How are a lever and a wheel and axle similar? **For more help, refer to the** Science Skill Handbook.

7. **Using Proportions** You are designing a lever to lift an object that weighs 500 N. The lever exerts the output force 1 m from the fulcrum. How far from the fulcrum must an effort force of 250 N be applied to lift the object? Draw a diagram as part of your answer. Label the forces and distances. **For more help, refer to the** Math Skill Handbook.

Activity

Design Your Own Experiment

Pulley Power

Imagine how long it might have taken to build the Sears Tower in Chicago without the aid of a pulley system attached to a crane. Hoisting the 1-ton I beams to a maximum height of 110 stories required large lifting forces and precise control of the beam's movement.

Construction workers also use smaller pulleys that are not attached to cranes to lift supplies to where they are needed. Pulleys are not limited to construction sites. They also are used to lift automobile engines out of cars, to help load and unload heavy objects on ships, and to lift heavy appliances and furniture.

Recognize the Problem

How can you use a pulley system to reduce the force needed to lift a load?

Form a Hypothesis

Write a hypothesis about how pulleys can be combined to make a system of pulleys to lift a heavy load, such as a brick. Consider the efficiency of your system.

Goals
- **Design** a pulley system.
- **Measure** the mechanical advantage and efficiency of the pulley system.

Possible Materials
single- and multiple-pulley systems
nylon rope
steel bar to support the pulley system
meterstick
*metric tape measure
variety of weights to test pulleys
force spring scale
brick
*heavy book
balance
*scale
*Alternate materials

Safety Precautions
The brick could be dangerous if it falls. Keep your hands and feet clear of it.

Test Your Hypothesis

Plan

1. Decide how you are going to support your pulley system. What materials will you use?

2. How will you measure the effort force and the resistance force? How will you determine the mechanical advantage? How will you measure efficiency?

3. **Experiment** by lifting small weights with a single pulley, double pulley, and so on. How efficient are the pulleys? In what ways can you increase the efficiency of your setup?

4. Use the results of step 3 to design a pulley system to lift the brick. Draw a diagram of your design. Label the different parts of the pulley system and use arrows to indicate the direction of movement for each section of rope.

Do

1. Make sure your teacher approves your plan before you start.

2. Assemble the pulley system you designed. You might want to test it with a smaller weight before attaching the brick.

3. **Measure** the force needed to lift the brick. How much rope must you pull to raise the brick 10 cm?

Analyze Your Data

1. **Calculate** the ideal mechanical advantage of your design.

2. **Calculate** the actual mechanical advantage of the pulley system you built.

3. **Calculate** the efficiency of your pulley system.

4. How did the mechanical advantage of your pulley system compare with those of your classmates?

Draw Conclusions

1. **Explain** how increasing the number of pulleys increases the mechanical advantage.

2. How could you modify the pulley system to lift a weight twice as heavy with the same effort force used here?

3. **Compare** this real machine with an ideal machine.

*C*ommunicating Your Data

Show your design diagram to the class. Review the design and point out good and bad characteristics of your pulley system. **For more help, refer to the** Science Skill Handbook.

Bionic Pe

People in need of transplants usually receive human organs. But many people's medical problems can only be solved by receiving artificial body parts. These synthetic devices, called prostheses, are used to replace anything from a heart valve to a knee joint. Bionics is the science of creating artificial body parts. A major focus of bionics is the replacement of lost limbs. Through accident, birth defect, or disease, people sometimes lack hands or feet, or even whole arms or legs. For centuries, people have used prostheses to replace limbs. In the past, disabled people used devices like peg legs or artificial arms that ended in a pair of hooks. These protheses didn't do much to replace lost functions of arms and legs.

But today, that's changed, thanks to the work of eighteenth-century scientists Luigi Galvani and Alessandro Volta. Because of their experiments, people began to realize that muscles contract by means of electrical impulses. This knowledge eventually led to an invention called functional neuromuscular stimulation (FNS). Some people are paralyzed because nerves that send electric impulses to certain muscles are destroyed.

FNS uses a computer or microprocessor to send electric impulses directly to these muscles. By sending the proper signals, the muscles can be made to move. FNS can allow paralyzed people to walk to a certain extent and to maintain muscle control.

The knowledge that muscles respond to electricity has helped create more effective prostheses. One such prostheses is the myoelectric arm. This battery-powered device connects muscle nerves in an amputated arm to a sensor.

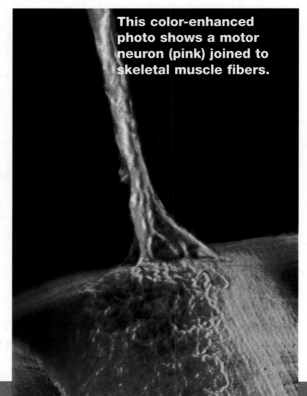

This color-enhanced photo shows a motor neuron (pink) joined to skeletal muscle fibers.

Artificial limbs can help people lead normal lives

ople

The sensor detects when the arm tenses, then transmits the signal to an artificial hand, which opens or closes. New prosthetic hands even give a sense of touch, as well as cold and heat.

Today's leg prostheses are also more sophisticated. The latest models are made of strong, lightweight titanium rods built inside a lifelike plastic covering. They allow users to compete in many sports, even tough ones like the triathlon. In addition, new artificial feet let wearers sense when their weight is on their toes, heels, or the sides of their feet. This gives them better balance.

A recent experiment promises even more amazing bionic technologies. Researchers have trained rats to move a robotic arm by using their brain signals that normally control movement—and without using their muscles at all. One day, this research might produce devices that allow paralyzed people to move artificial limbs just by brain power.

People with leg prostheses can participate in sports.

Myoelectric arms make life easier for people who have them.

CONNECTIONS Research Use your school's media center to find other aspects of robotics such as walking machines or robots that perform planetary exploration. What are they used for? How do they work? You could take it one step further and learn about cyborgs. Report to the class.

SCIENCE *Online*

For more information, visit science.glencoe.com

Reviewing Main Ideas

Section 1 Work and Power

1. Work is done when a force exerted on an object causes the object to move.

2. A force can do work only when it is exerted in the same direction as the object moves. *Is work being done if this car is stuck?*

3. Work is equal to force times distance, and the unit of work is the joule.

4. Power is the rate at which work is done, and the unit of power is the watt.

Section 2 Using Machines

1. A machine changes the size or direction of the input force or the distance over which it is exerted.

2. The mechanical advantage of a machine is its output force divided by its input force. *What is the mechanical advantage of this machine?*

Section 3 Simple Machines

1. A machine that does work with only one movement is a simple machine. A compound machine is a combination of simple machines.

2. Simple machines include the inclined plane, lever, wheel and axle, screw, wedge, and pulley. *What type of simple machine is shown?*

3. Wedges and screws are two types of inclined planes. The mechanical advantage of an inclined plane is its length divided by its height.

4. The mechanical advantage of a lever depends on the location of the fulcrum. The mechanical advantage of a wheel and axle depends on the radius of each circular object.

5. Pulleys can be used to multiply force and change direction. The mechanical advantage of a fixed pulley is one and of a single movable pulley, two.

FOLDABLES
Reading & Study Skills

After You Read

To help you review work and simple machines, use the Foldable you made at the beginning of the chapter.

Visualizing Main Ideas

Complete the following concept map on simple machines.

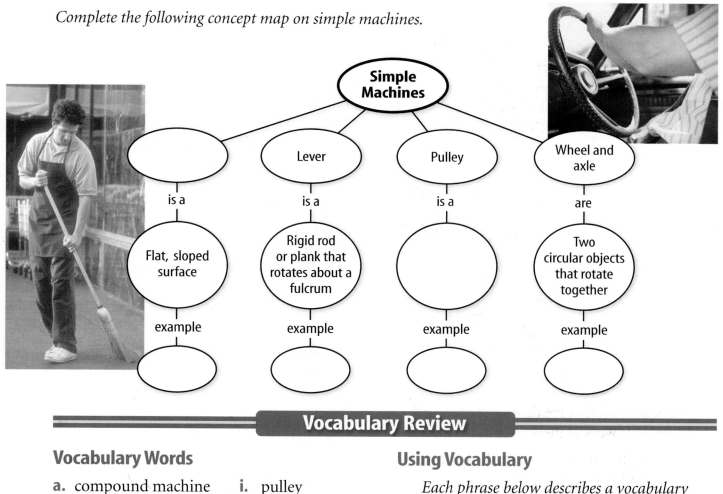

Vocabulary Review

Vocabulary Words

a. compound machine
b. efficiency
c. inclined plane
d. input force
e. lever
f. mechanical advantage
g. output force
h. power

i. pulley
j. screw
k. simple machine
l. wedge
m. wheel and axle
n. work

Using Vocabulary

Each phrase below describes a vocabulary word. Write the vocabulary word that matches the phrase describing it.

1. percentage of work in to work out
2. force put into a machine
3. force exerted on an object by a machine
4. two rigidly attached wheels
5. input force divided by output force
6. a machine with only one movement
7. an inclined plane that moves
8. a rigid rod that rotates about a fulcrum
9. a flat, sloped surface
10. amount of work divided by time

THE PRINCETON REVIEW Study Tip

Without looking back at your textbook, write a summary of each section of the chapter after you've read it. If you write it in your own words, you will better remember it.

Choose the word or phrase that best answers the question.

1. Which of the following is a requirement for work to be done?
   A) Force is exerted.   C) Force moves object.
   B) Object is carried.   D) Machine is used.

2. How much work is done when a force of 30 N moves an object a distance of 3 m?
   A) 3 J   C) 30 J
   B) 10 J   D) 90 J

3. How much power is expended when 600 J of work are done in 10 s?
   A) 6 W   C) 600 W
   B) 60 W   D) 610 W

4. Which is an example of a simple machine?
   A) baseball bat   C) can opener
   B) bicycle   D) car

5. What is mechanical advantage?
   A) input force/output force
   B) output force/input force
   C) input work/output work
   D) output work/input work

6. What is the ideal mechanical advantage of a machine that changes only the direction of the input force?
   A) less than 1   C) 1
   B) zero   D) greater than 1

7. A wheel with a radius of 20 cm is attached to an axle with a radius of 1 cm. What is the output force if the input force on the wheel is 100 N?
   A) 5 N   C) 500 N
   B) 200 N   D) 2,000 N

8. Which of the following is a form of the inclined plane?
   A) pulley   C) wheel and axle
   B) screw   D) lever

9. A ramp decreases which of the following?
   A) height   C) input force
   B) output force   D) input distance

10. If a machine takes in 50 J and puts out 45 J, what is its efficiency?
    A) 0.9 percent   C) 90 percent
    B) 1.1 percent   D) 111 percent

11. Explain why the output work for any machine can't be greater than the input work.

12. A doorknob is an example of a wheel and axle. Explain why turning the knob is easier than turning the axle.

13. What is the mechanical advantage of a 6-m long ramp that extends from a ground-level sidewalk to a 2-m high porch?

14. How much input force is required to lift an 11,000-N beam using a pulley system with a mechanical advantage of 20?

15. Would a 9 N force applied 2 m from the fulcrum lift the weight? Explain.

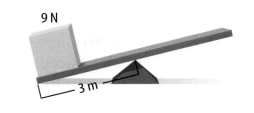

9 N

3 m

16. **Measuring in SI** At the 1976 Olympics, Vasili Alexeyev shattered the world record for weight lifting when he lifted 2,500 N from the floor to a point over his head 2 m above the ground. It took him about 5 s to complete the lift. How much work did he do? What was his power?

17. Predicting Suppose a lever is in balance. Would this arrangement be in balance on the Moon, where the force of gravity is less? Explain.

18. Making and Using Graphs A pulley system has a mechanical advantage of 5. Make a graph of the possible combinations of input force and output force.

19. Solving One-Step Equations If you put 8,000 J of work into a machine with an efficiency of 60 percent, what is the work output?

20. Recognizing Cause and Effect The diagram below shows a force exerted at an angle to pull a sled. How much work is done if the sled moves 10 m horizontally?

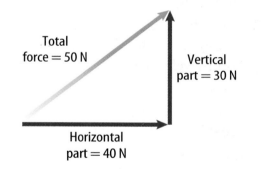

Total force = 50 N

Vertical part = 30 N

Horizontal part = 40 N

Performance Assessment

21. Identifying Levers You have levers in your body. Your muscles and tendons provide the input force. Your joints act as fulcrums. The output force is used to move everything from your head to your hands. Describe and draw any human levers you can identify.

TECHNOLOGY

Go to the Glencoe Science Web site at **science.glencoe.com** or use the **Glencoe Science CD-ROM** for additional chapter assessment.

THE PRINCETON REVIEW   Test Practice

At state fairs, participants often compete in pulling a 250-pound rock to a 50-meter height by using a simple pulley system. Each participant gets three tries to reach the 50-meter height.

Top Four Competitors				
Participant	Bud	Charley	Maxine	Joey
Attempt 1	35 m	18 m	38 m	43 m
Attempt 2	28 m	23 m	35 m	38 m
Attempt 3	42 m	30 m	36 m	21 m

Study the table and answer the following questions.

1. According to this information, which of the participants pulled the 250-pound rock to the highest average height?
A) Bud   **C)** Maxine
B) Charley   **D)** Joey

2. According to the table, the competitor who did the least amount of total work overall was _____ .
F) Bud   **H)** Maxine
G) Charley   **J)** Joey

3. According to the table, which of the following is true about the amount of work done by Joey over his three attempts, compared to the other top four competitors?
A) The range is the smallest
B) The range is the largest
C) The average is the lowest
D) The average is the highest

Energy

Volcanoes, earthquakes, light-ning, and hurricanes produce some of the most powerful forces in nature. Every one of these phenomena contains a tremendous amount of energy. The river of lava shown in this picture flowing from Mount Etna in Italy has heat energy, light energy, and energy of motion. In this chapter, you will learn about different forms and sources of energy. You also will learn how energy can be transformed from one form into another, and how some forms of energy can be used.

What do you think?

Science Journal Look at the picture below with a classmate. Discuss what is happening. Here's a hint: *Concentrating energy is the key to what is happening here.* Write your answer or best guess in your Science Journal.

EXPLORE ACTIVITY

A marble and a piece of wood are on a countertop. If nothing disturbs them, they will remain there. However, if you tilt the wood and roll the marble down the slope, the marble acquires a new property—the ability to do something.

Analyze a marble launch

1. Make a track by slightly separating two metersticks placed side by side.
2. On a table, raise one end of the track slightly and measure the height.
3. Roll a marble down the track. Measure the distance from its starting point to where it hits the floor. Repeat. Calculate the average of the two measurements.
4. Repeat steps 2 and 3 for three different heights. Predict what will happen if you use a heavier marble. Test your prediction and record your observations.

Observe

In your Science Journal, describe your experiment and what you discovered. How did the different heights cause the distance to change?

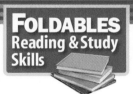

Before You Read

FOLDABLES
Reading & Study Skills

Making a Know-Want-Learn Study Fold Make the following Foldable to help identify what you already know and what you want to know about energy.

1. Place a sheet of paper in front of you so the long side is at the top. Fold the paper in half from top to bottom.
2. Fold both sides in. Unfold the paper so three sections show.
3. Through the top thickness of paper, cut along each of the fold lines to the topfold, forming three tabs. Label the tabs *Know*, *Want*, and *Learned*.
4. Before you read the chapter, write what you know and what you want to know under the tabs. As you read the chapter, correct what you have written and add more questions.

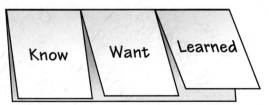

What is energy?

As You Read

What You'll Learn

- **Explain** what energy is.
- **Distinguish** between kinetic energy and potential energy.
- **Identify** the various forms of energy.

Vocabulary

energy
kinetic energy
potential energy
thermal energy

chemical energy
radiant energy
electrical energy
nuclear energy

Why It's Important

Energy is the source of all activity.

Figure 1
Energy is the ability to cause change. *How can these objects cause change?*

The Nature of Energy

What comes to mind when you hear the word *energy?* Do you picture running, leaping, and spinning like a dancer or a gymnast? How would you define energy? When an object has energy, it can make things happen. In other words, **energy** is the ability to cause change. What do the items shown in **Figure 1** have in common?

Look around and notice the changes that are occurring—someone walking by or a ray of sunshine that is streaming through the window and warming your desk. Maybe you can see the wind moving the leaves on a tree. What changes are occurring?

Transferring Energy You might not realize it, but you have a large amount of energy. In fact, everything around you has energy, but you notice it only when a change takes place. Anytime a change occurs, energy is transferred from one object to another. You hear a footstep because energy is transferred from a foot hitting the ground to your ears. Leaves are put into motion when energy in the moving wind is transferred to them. The spot on the desktop becomes warmer when energy is transferred to it from the sunlight. In fact, all objects, including leaves and desktops, have energy.

Energy of Motion

Things that move can cause change. A bowling ball rolls down the alley and knocks down some pins, as in **Figure 2A.** Is energy involved? A change occurs when the pins fall over. The bowling ball causes this change, so the bowling ball has energy. The energy in the motion of the bowling ball causes the pins to fall. As the ball moves, it has a form of energy called kinetic energy. **Kinetic energy** is the energy an object has due to its motion. If an object isn't moving, it doesn't have kinetic energy.

Kinetic Energy and Speed If you roll the bowling ball so it moves faster, what happens when it hits the pins? It might knock down more pins, or it might cause the pins to go flying farther. A faster ball causes more change to occur than a ball that is moving slowly. Look at **Figure 2B.** The professional bowler rolls a fast-moving bowling ball. When her ball hits the pins, pins go flying faster and farther than for a slower-moving ball. All that action signals that her ball has more energy. The faster the ball goes, the more kinetic energy it has. This is true for all moving objects. Kinetic energy increases as an object moves faster.

Kinetic Energy and Mass Suppose, as shown in **Figure 2C,** you roll a volleyball down the alley instead of a bowling ball. If the volleyball travels at the same speed as a bowling ball, do you think it will send pins flying as far? The answer is no. The volleyball might not knock down any pins. Does the volleyball have less energy than the bowling ball even though they are traveling at the same speed? An important difference between the volleyball and the bowling ball is that the volleyball has less mass. Even though the volleyball is moving at the same speed as the bowling ball, the volleyball has less kinetic energy because it has less mass. Kinetic energy also depends on the mass of a moving object. Kinetic energy increases as the mass of the object increases.

 **Reading Check** *Why does a volleyball knock over fewer pins than a bowling ball?*

Figure 2
The kinetic energy of an object depends on two quantities.
What are those quantities?

A This ball has kinetic energy because it is rolling down the alley.

B This ball has more kinetic energy because it has more speed.

C This ball has less kinetic energy because it has less mass.

Figure 3
The potential energy of an object depends on its mass and height above the ground. *Which vase has more potential energy, the red one or the blue one?*

Energy of Position

An object can have energy even though it is not moving. For example, a glass of water sitting on the kitchen table doesn't have any kinetic energy because it isn't moving. If you accidentally nudge the glass and it falls on the floor, changes occur. Gravity pulls the glass downward, and the glass has energy of motion as it falls. Where did this energy come from?

When the glass was sitting on the table, it had potential (puh TEN chul) energy. **Potential energy** is the energy stored in an object because of its position. In this case, the position is the height of the glass above the floor. The potential energy of the glass changes to kinetic energy as the glass falls. The potential energy of the glass is greater if it is higher above the floor. Potential energy also depends on mass. The more mass an object has, the more potential energy it has. Which object in **Figure 3** has the most potential energy?

Forms of Energy

Food, sunlight, and wind have energy, yet they seem different because they contain different forms of energy. Food and sunlight contain forms of energy different from the kinetic energy in the motion of the wind. The warmth you feel from sunlight is another type of energy that is different from the energy of motion or position.

Thermal Energy The feeling of warmth from sunlight signals that your body is acquiring more thermal energy. All objects have **thermal energy** that increases as its temperature increases. A cup of hot chocolate has more thermal energy than a cup of cold water, as shown in **Figure 4.** Similarly, the cup of water has more thermal energy than a block of ice of the same mass. Your body continually produces thermal energy. Many chemical reactions that take place inside your cells produce thermal energy. Where does this energy come from? Thermal energy released by chemical reactions comes from another form of energy called chemical energy.

Figure 4
The hotter an object is, the more thermal energy it has. A cup of hot chocolate has more thermal energy than a cup of water, which has more thermal energy than a block of ice with the same mass.

Figure 5
Complex chemicals in food store chemical energy. During activity, the chemical energy transforms into kinetic energy.

Sugar

Chemical Energy When you eat a meal, you are putting a source of energy inside your body. Food contains chemical energy that your body uses to provide energy for your brain, to power your movements, and to fuel your growth. As in **Figure 5,** food contains chemicals, such as sugar, which can be broken down in your body. These chemicals are made of atoms that are bonded together, and energy is stored in the bonds between atoms. **Chemical energy** is the energy stored in chemical bonds. When chemicals are broken apart and new chemicals are formed, some of this energy is released. The flame of a candle is the result of chemical energy stored in the wax. When the wax burns, chemical energy is transformed into thermal energy and light energy.

Light Energy Light from the candle flame travels through the air at an incredibly fast speed of 300,000 km/s. This is fast enough to circle Earth almost eight times in 1 s. When light strikes something, it can be absorbed, transmitted, or reflected. If the light is absorbed, it will cause the object to warm up. In other words, the thermal energy of the object has increased because light transferred energy to it. The type of energy light carries is called **radiant energy. Figure 6** shows a coil of wire that produces radiant energy when it is heated. To heat the metal, another type of energy can be used—electrical energy.

Figure 6
Electrical energy is transformed into thermal energy in the metal heating coil. As the metal becomes hotter, it emits more radiant energy.

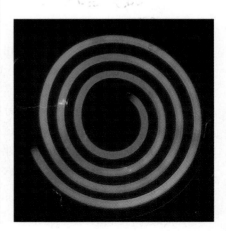

✓ Reading Check *How do you know that light has energy?*

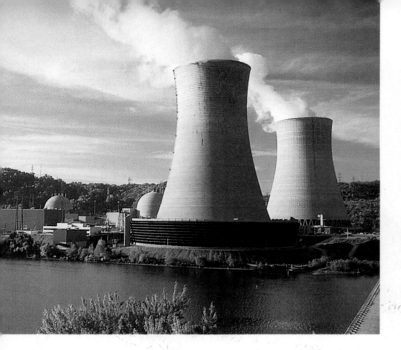

Electrical Energy Electrical lighting is one of the many ways electrical energy is used. Look around at all the devices that use electrical energy. The electric current that comes out of batteries and wall sockets carries **electrical energy.** The amount of electrical energy depends on the voltage. The current out of a 120-V wall socket can carry more energy than the current out of a 1.5-V battery. To produce the enormous quantities of electrical energy consumed each day, large power plants are needed. In the United States, about 20 percent of the electrical energy that is generated comes from nuclear power plants.

Figure 7
Complex power plants are required to obtain useful energy from the nucleus of an atom.

Nuclear Energy Nuclear power plants use the energy stored in the nucleus of an atom to generate electricity. Every atomic nucleus contains energy—**nuclear energy**—that can be transformed into other forms of energy. However, releasing the nuclear energy is a difficult process. It involves the construction of complex power plants, as shown in **Figure 7.** In contrast, all that is needed to release chemical energy from wood is a match.

Section 1 Assessment

1. How do you know if an object has energy? Do you have energy? Does a rock?

2. Contrast chemical and nuclear energy.

3. How can chemical energy transform into thermal energy? Into light energy?

4. If two vases are side by side on a high shelf, could one have more potential energy than the other? Explain.

5. **Think Critically** A golf ball and a bowling ball have the same kinetic energy. Which one is moving faster? Explain your answer using what you know about kinetic energy. Suppose the golf ball and the bowling ball have the same speed. Which of the two has more kinetic energy?

Skill Builder Activities

6. **Interpreting Data** Review your results from the Explore Activity. Where did the marble have the most kinetic energy? Where did the marble have the most potential energy? Can you infer a relationship between kinetic energy and potential energy based on your observations? **For more help, refer to the** Science Skill Handbook.

7. **Communicating** The term *energy* is used in everyday language. In your Science Journal, record different expressions and ways of using the word *energy.* Decide which ones match the definition of energy presented in this section. **For more help, refer to the** Science Skill Handbook.

Energy Transformations

Changing Forms of Energy

Chemical, thermal, radiant, and electrical are some of the forms that energy can have. In the world around you, energy is transforming continually between one form and another. You observe some of these transformations by noticing a change in your environment. Forest fires are a dramatic example of an environmental change that can occur naturally as a result of lightning strikes. Another type of change, shown in **Figure 8,** is a mountain biker pedaling to the top of a hill. What energy transformations occur as he moves up the hill?

Tracking Energy Transformations As the mountain biker pedals, many energy transformations are taking place. In his leg muscles, chemical energy is transforming into kinetic energy. The kinetic energy of his leg muscle transforms into kinetic energy of the bicycle. Some of this energy transforms into potential energy as he moves up the hill. Also, some energy is transformed into thermal energy. His body is warmer because chemical energy is being released. Because of friction, the mechanical parts of the bicycle are warmer, too.

As You Read

What **You'll Learn**
- **Apply** the law of conservation of energy to energy transformations.
- **Identify** how energy changes form.
- **Describe** how electric power plants produce energy.

Vocabulary
law of conservation of energy
generator
turbine

Why **It's Important**
Many devices you use every day change energy from one form to another.

Figure 8
The ability to transform energy allows the biker to climb the hill.
Identify all the forms of energy that are represented in the photograph.

The Law of Conservation of Energy

It can be a challenge to track energy as it moves from object to object. However, one extremely important principle can serve as a guide as you trace the flow of energy. According to the **law of conservation of energy,** energy is never created or destroyed. The only thing that changes is the form in which energy appears. When the biker is resting at the summit, all his original energy is still around. Some of the energy is in the form of potential energy, which he will use as he coasts down the hill. Some of this energy was changed to thermal energy by friction in the bike. Chemical energy was also changed to thermal energy in the biker's muscles, making him feel hot. As he rests, this thermal energy moves from his body to the air around him. No energy is missing—it can all be accounted for.

✓ **Reading Check** *Can energy ever be lost? Why or why not?*

Changing Kinetic and Potential Energy

The law of conservation of energy can be used to identify the energy changes in a system, especially if the system is not too complicated. For example, tossing a ball into the air and catching it is a simple system. As shown in **Figure 9,** as the ball leaves your hand, most of its energy is kinetic. As the ball rises, it slows and loses kinetic energy. But, the total energy of the ball hasn't changed. The loss of kinetic energy equals the gain of potential energy as the ball flies higher in the air. The total amount of energy always remains constant. Energy moves from place to place and changes form, but it never is created or destroyed.

Research Visit the Glencoe Science Web site at **science.glencoe.com** for more information about how energy changes form when it is transformed from one form to another. Use a spreadsheet program to summarize what you've learned.

Figure 9
During the flight of the baseball, energy is transforming between kinetic and potential energy.
Where does the ball have the most kinetic energy? Where does the ball have the most total energy?

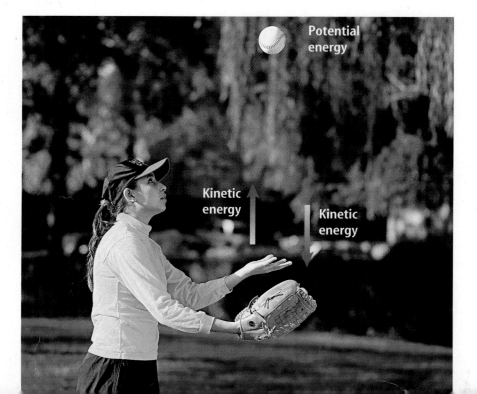

Figure 10

Hybrid cars that use an electric motor and a gasoline engine for power are now available. These cars get up to 29 km/L. Inventions such as the hybrid car make energy transformations more efficient.

Battery

Gasoline engine

Generator

Electric motor

Energy Changes Form

Energy transformations occur constantly all around you. Many machines are devices that transform energy from one form to another. For example, an automobile engine transforms the chemical energy in gasoline into energy of motion. However, not all of the chemical energy is converted into kinetic energy. Instead, some of the chemical energy is converted into thermal energy, and the engine becomes hot. An engine that converts chemical energy into more kinetic energy is a more efficient engine. New types of cars, like the one shown in **Figure 10,** use an electric motor along with a gasoline engine. These engines are more efficient so the car can travel farther on a gallon of gas.

Life Science
INTEGRATION

Transforming Chemical Energy

Inside your body, chemical energy also is transformed into kinetic energy. Look at **Figure 11.** The transformation of chemical to kinetic energy occurs in muscle cells. There, chemical reactions take place that cause certain molecules to change shape. Your muscle contracts when many of these changes occur, and a part of your body moves.

The matter contained in living organisms, or biomass, contains chemical energy. When organisms die, chemical compounds in their biomass break down, or decompose. Bacteria, fungi, and other organisms help convert these chemical compounds to simpler chemicals that can be used by other living things.

Thermal energy also is released as these changes occur. For example, a compost pile can contain plant matter, such as grass clippings and leaves. As the compost pile decomposes, chemical energy is converted into thermal energy. This can cause the temperature of a compost pile to reach 60°C.

TRY AT HOME
Mini LAB

Analyzing Energy Transformations

Procedure 🥽 👕

1. Place soft **clay** on the floor and smooth out its surface.
2. Hold a **marble** 1.5 m above the clay and drop it. Measure the depth of the crater made by the marble.
3. Repeat this procedure using a **steel ball**, a **rubber ball**, and a **table-tennis ball**.

Analysis

1. Compare the depths of the craters to determine which ball had the most kinetic energy as it hit the clay. Why did this ball have the most kinetic energy?
2. Explain how potential energy was transformed into kinetic energy during your activity.

Figure 11

Paddling a raft, throwing a baseball, playing the violin — your skeletal muscles make these and countless other body movements possible. Muscles work by pulling, or contracting. At the cellular level, muscle contractions are powered by reactions that transform chemical energy into kinetic energy.

▶ Energy transformations taking place in your muscles provide the power to move.

▲ Many skeletal muscles are arranged in pairs that work in opposition to each other. When you bend your arm, the biceps muscle contracts, while the triceps relaxes. When you extend your arm the triceps contracts, and the biceps relaxes.

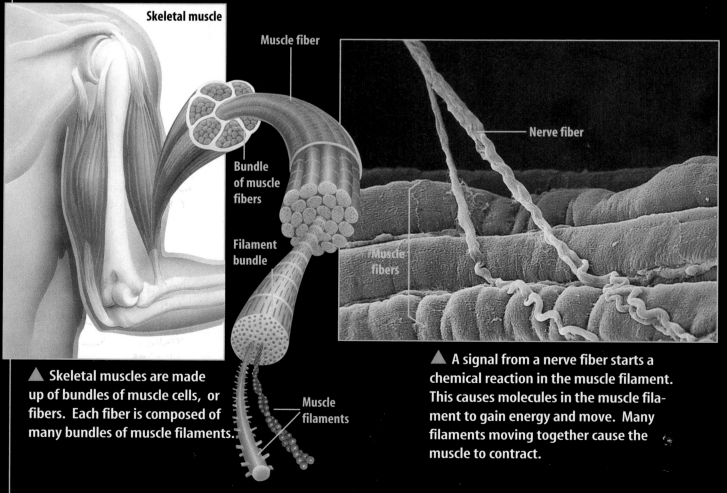

Skeletal muscle

Muscle fiber

Bundle of muscle fibers

Filament bundle

Muscle filaments

Nerve fiber

Muscle fibers

▲ Skeletal muscles are made up of bundles of muscle cells, or fibers. Each fiber is composed of many bundles of muscle filaments.

▲ A signal from a nerve fiber starts a chemical reaction in the muscle filament. This causes molecules in the muscle filament to gain energy and move. Many filaments moving together cause the muscle to contract.

Figure 12
The simple act of listening to a radio involves many energy transformations. A few are diagrammed here.

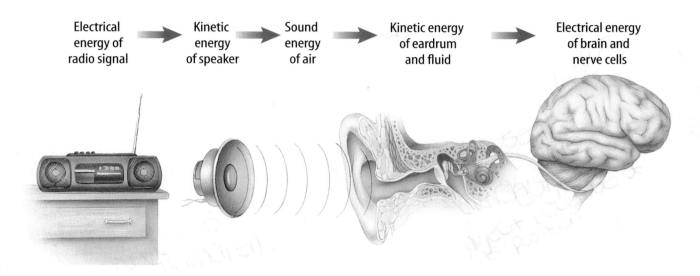

Electrical energy of radio signal → Kinetic energy of speaker → Sound energy of air → Kinetic energy of eardrum and fluid → Electrical energy of brain and nerve cells

Transforming Electrical Energy Every day you use electrical energy. When you flip a light switch, or turn on a radio or television, or use a hair drier, you are transforming electrical energy to other forms of energy. Every time you plug something into a wall outlet, or use a battery, you are using electrical energy. **Figure 12** shows how electrical energy is transformed into other forms of energy when you listen to a radio. A loudspeaker in the radio converts electrical energy into sound waves that travel to your ear—energy in motion. The energy that is carried by the sound waves causes parts of the ear to move also. This energy of motion is transformed again into chemical and electrical energy in nerve cells, which send the energy to your brain. After your brain interprets this energy as a voice or music, where does the energy go? The energy finally is transformed into thermal energy.

Transforming Thermal Energy Different forms of energy can be transformed into thermal energy. For example, chemical energy changes into thermal energy when something burns. Electrical energy changes into thermal energy when a wire that is carrying an electric current gets hot. Thermal energy can be used to heat buildings and keep you warm. Thermal energy also can be used to heat water. If water is heated to its boiling point, it changes to steam. This steam can be transformed to kinetic energy by steam engines, like the steam locomotives that used to pull trains. Thermal energy also can be transformed into radiant energy. For example, when a bar of metal is heated to a high temperature, it glows and gives off light.

Life Science
INTEGRATION

Most organisms have some adaptation for controlling the amount of thermal energy in their bodies. Some living in cooler climates have thick fur coats that help prevent thermal energy from escaping, and some living in desert regions have skin that helps keep thermal energy out. Research some of the adaptations different organisms have for controlling the thermal energy in their bodies.

How Thermal Energy Moves Thermal energy can move from one place to another. Look at **Figure 13.** The hot chocolate has thermal energy that moves from the cup to the cooler air around it, and to the cooler spoon. Thermal energy only moves from something at a higher temperature to something at a lower temperature.

Thermal energy

Figure 13
Thermal energy moves from the hot chocolate to the cooler surroundings. *What happens to the hot chocolate as it loses thermal energy?*

Generating Electrical Energy

The enormous amount of electrical energy that is used every day is too large to be stored in batteries. The electrical energy that is available for use at any wall socket must be generated continually by power plants. Every power plant works on the same principle—energy is used to turn a large generator. A **generator** is a device that transforms kinetic energy into electrical energy. In fossil fuel power plants, coal, oil, or natural gas is burned to boil water. As the hot water boils, the steam rushes through a **turbine,** which contains a set of narrowly spaced fan blades. The steam pushes on the blades and turns the turbine, which in turn rotates a shaft in the generator to produce the electrical energy, as shown in **Figure 14.**

Figure 14
A coal-burning power plant transforms the chemical energy in coal into electrical energy. *What are some of the other energy sources that power plants use?*

✓ **Reading Check** *What does a generator do?*

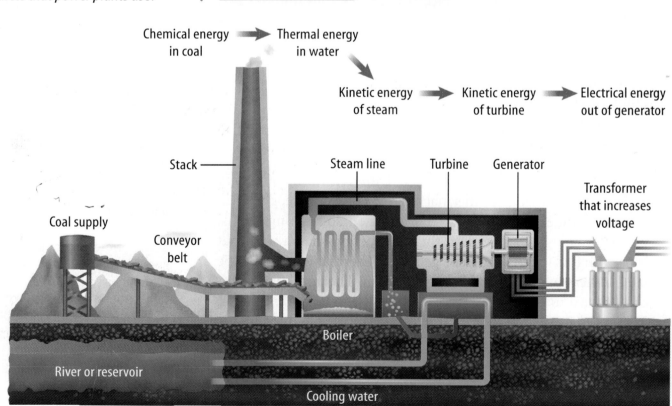

Chemical energy in coal → Thermal energy in water → Kinetic energy of steam → Kinetic energy of turbine → Electrical energy out of generator

Stack

Steam line   Turbine   Generator

Transformer that increases voltage

Coal supply

Conveyor belt

Boiler

River or reservoir

Cooling water

Power Plants Almost 90 percent of the electrical energy generated in the United States is produced by nuclear and fossil fuel power plants, as shown in **Figure 15.** Other types of power plants include hydroelectric (hi droh ih LEK trihk) and wind. Hydroelectric power plants transform the kinetic energy of moving water into electrical energy. Wind power plants transform the kinetic energy of moving air into electrical energy. In these power plants, a generator converts the kinetic energy of moving water or wind to electrical energy.

To analyze the energy transformations in a power plant, you can diagram the energy changes using arrows. A coal-burning power plant generates electrical energy through the following series of energy transformations.

chemical energy of coal	$\rightarrow$	thermal energy of water	$\rightarrow$	kinetic energy of steam	$\rightarrow$	kinetic energy of turbine	$\rightarrow$	electrical energy out of generator

Nuclear power plants use a similar series of transformations. Hydroelectric plants, however, skip the steps that change water into steam because the water strikes the turbine directly.

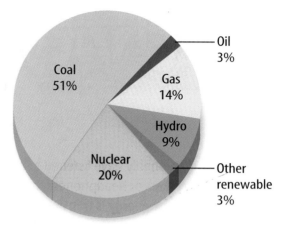

Figure 15
The graph shows sources of electrical energy in the United States. *Which energy source do you think is being used to provide the electricity for the lights overhead?*

Section 2 Assessment

1. What is the law of conservation of energy?

2. If your body temperature is 37°C and you are sitting in a room which has a temperature of 25°C, does your body gain or lose thermal energy? Explain.

3. What are the basic steps involved in generating electricity at a power plant?

4. Diagram the energy transformations that take place at a hydroelectric power plant.

5. **Think Critically** You begin pedaling your bicycle, making it move faster and faster. You notice that at first it is easy to speed up, but then it becomes difficult. You pedal with all your strength, yet you cannot go any faster. Use energy concepts to explain what is happening.

Skill Builder Activities

6. **Testing a Hypothesis** If you drop a rubber ball onto a hard surface, the first bounce will be the highest. How much lower will the second bounce be? If you drop the ball on the top of a shoe box, will it bounce as high? Make a hypothesis. Design and conduct an experiment to test your hypothesis. **For more help, refer to the** Science Skill Handbook.

7. **Using Graphics Software** Use graphics software to diagram all the energy transformations that take place during a conversation. What forms of energy are in the sequence from one person making a sound to a second person hearing that sound? **For more help, refer to the** Technology Skill Handbook.

Activity

Hearing with Your Jaw

You probably have listened to music using speakers or headphones. Have you ever considered how energy is transferred to get the energy from the radio or CD player to your brain? What type of energy is needed to power the radio or CD player? Where does this energy come from? How does that energy become sound? How does the sound get to you? In this activity, the sound from a radio or CD player is going to travel through a motor before entering your body through your jaw instead of your ears.

What You'll Investigate
How can energy be transferred from a radio or CD player to your brain?

Materials
radio or CD player
small electrical motor
headphone jack

Goals
■ **Identify** energy transfers and transformations.
■ **Explain** your results in terms of transformations of energy and conservation of energy.

Safety Symbols

Procedure

1. Go to one of the places in the room with a motor/radio assembly.

2. Turn on the radio or CD player so that you hear the music.

3. Push the headphone jack into the headphone plug on the radio or CD player.

4. Press the axle of the motor against the side of your jaw.

Conclude and Apply

1. **Describe** what you heard in your Science Journal.

2. What type of energy did you have in the beginning? In the end?

3. **Draw** a diagram to show all of the energy transformations taking place.

4. Did anything get hotter as a result of this activity? Explain.

5. **Explain** your results using the law of conservation of energy.

Communicating Your Data

Compare your conclusions with those of other students in your class. **For more help, refer to the** Science Skill Handbook.

Sources of Energy

Using Energy

Press a button on the remote control and your favorite program appears on television. Open your refrigerator and pull out something cold to drink. Ride to the mall in a car. For any of these things to occur, a transfer of energy must take place. Radiant energy is transferred to your television, electrical energy is transferred to your refrigerator, and the chemical energy in gasoline is transferred to the engine of the car.

Every day energy is used to provide light and to heat and cool homes, schools, and workplaces. Energy is used to run cars, buses, trucks, trains, and airplanes that transport people and materials from one place to another. Energy also is used to make clothing and other materials and to cook food.

According to the law of conservation of energy, energy can't be created or destroyed. Energy only can change form. If a car or refrigerator can't create the energy they use, then where does this energy come from?

Energy Resources

Energy cannot be made, but must come from the natural world. As you can see in **Figure 16,** the surface of Earth receives energy from two sources—the Sun and radioactive atoms in Earth's interior. Of these two energy sources, the energy from the Sun has much more impact on your life. Nearly all the energy you used today can be traced to the Sun, even the gasoline used to power the car or school bus you came to school in.

As You Read

What **You'll Learn**

- **Explain** what renewable, nonrenewable, and alternative resources are.
- **Describe** the advantages and disadvantages of using various energy sources.

Vocabulary
nonrenewable resource
renewable resource
alternative resource
inexhaustible resource
photovoltaic

Why **It's Important**

Energy is vital for survival and making life comfortable. Developing new energy sources will improve modern standards of living.

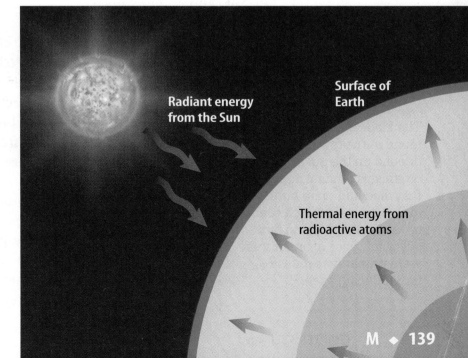

Radiant energy from the Sun

Surface of Earth

Thermal energy from radioactive atoms

Figure 16
All the energy you use can be traced to one of two sources—the Sun or radioactive atoms in Earth's interior.

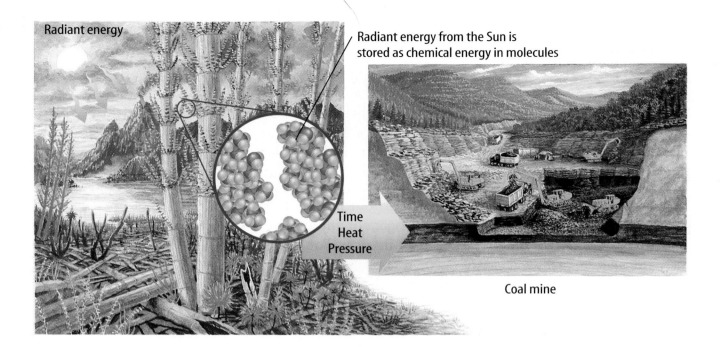

Radiant energy

Radiant energy from the Sun is stored as chemical energy in molecules

Time
Heat
Pressure

Coal mine

Figure 17
Coal is formed after the molecules in ancient plants are heated under pressure for millions of years. The energy stored by the molecules in coal originally came from the Sun.

Earth Science

INTEGRATION

The kinds of fossil fuels found in the ground depend on the kinds of organisms (animal or plant) that died and were buried in that spot. Research coal, oil, and natural gas to find out what types of organisms were primarily responsible for producing each.

Fossil Fuels

Fossil fuels are coal, oil, and natural gas. Oil and natural gas were made from the remains of microscopic organisms that lived in Earth's oceans millions of years ago. Heat and pressure gradually turned these ancient organisms into oil and natural gas. Coal was formed by similar process from the remains of ancient plants that once lived on land, as shown in **Figure 17.**

Through the process of photosynthesis, ancient plants converted the radiant energy in sunlight to chemical energy stored in various types of molecules. Heat and pressure changed these molecules into other types of molecules as fossil fuels formed. Chemical energy stored in these molecules is released when fossil fuels are burned.

Using Fossil Fuels The energy used when you ride in a car, turn on a light, or use an electric appliance usually comes from burning fossil fuels. However, it takes millions of years to replace each drop of gasoline and each lump of coal that is burned. This means that the supply of oil on Earth will continue to decrease as oil is used. An energy source that is used up much faster than it can be replaced is a **nonrenewable resource.** Fossil fuels are nonrenewable resources.

Burning fossil fuels to produce energy also generates chemical compounds that cause pollution. Each year billions of kilograms of air pollutants are produced by burning fossil fuels. These pollutants can cause respiratory illnesses and acid rain. Also, the carbon dioxide gas formed when fossil fuels are burned might cause Earth's climate to warm.

Nuclear Energy

Can you imagine running an automobile on 1 kg of fuel that releases almost 3 million times more energy than 1 L of gas? What could supply so much energy from so little mass? The answer is the nuclei of uranium atoms. Some of these nuclei are unstable and break apart, releasing enormous amounts of energy in the process. This energy can be used to generate electricity by heating water to produce steam that spins an electric generator, as shown in **Figure 18.** Because no fossil fuels are burned, generating electricity using nuclear energy helps make the supply of fossil fuels last longer. Also, unlike fossil fuel power plants, nuclear power plants produce almost no air pollution. In one year, a typical nuclear power plant generates enough energy to supply 600,000 homes with power and produces only 1 m^3 of waste.

Nuclear Wastes Like all energy sources, nuclear energy has its advantages and disadvantages. One disadvantage is the amount of uranium in Earth's crust is nonrenewable. Another is that the waste produced by nuclear power plants is radioactive and can be dangerous to living things. Some of the materials in the nuclear waste will remain radioactive for many thousands of years. As a result the waste must be stored so no radioactivity is released into the environment for a long time. One method is to seal the waste in a ceramic material, place the ceramic in protective containers, and then bury the containers far underground. However, the burial site would have to be chosen carefully so underground water supplies aren't contaminated. Also, the site would have to be safe from earthquakes and other natural disasters that might cause radioactive material to be released.

Figure 18
To obtain electrical energy from nuclear energy, a series of energy transformations must occur.

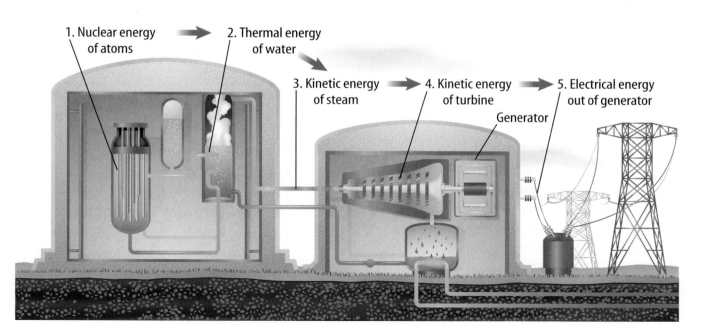

1. Nuclear energy of atoms → 2. Thermal energy of water → 3. Kinetic energy of steam → 4. Kinetic energy of turbine → 5. Electrical energy out of generator

Generator

Hydroelectricity

Currently, transforming the potential energy of water that is trapped behind dams supplies the world with almost 20 percent of its electrical energy. Hydroelectricity is the largest renewable source of energy. A **renewable resource** is an energy source that is replenished continually. As long as enough rain and snow fall to keep rivers flowing, hydroelectric power plants can generate electrical energy, as shown in **Figure 19.**

Although production of hydroelectricity is largely pollution free, it has one major problem. It disrupts the life cycle of aquatic animals, especially fish. This is particularly true in the Northwest where salmon spawn and run. Because salmon return to the spot where they were hatched to lay their eggs, the development of dams has hindered a large fraction of salmon from reproducing. This has greatly reduced the salmon population. Efforts to correct the problem have resulted in plans to remove a number of dams. In an attempt to help fish bypass some dams, fish ladders are being installed. Like most energy sources, hydroelectricity has advantages and disadvantages.

SCIENCE *Online*

Data Update Visit the Glencoe Science Web site at **science.glencoe.com** for data about the use of hydroelectricity in various parts of the world. Using a classroom map, present your finds to your class.

Problem-Solving Activity

Is energy consumption outpacing production?

You use energy every day—to get to school, to watch TV, and to heat or cool your home. The amount of energy consumed by an average person has increased over the last 50 years. Consequently, more energy must be produced.

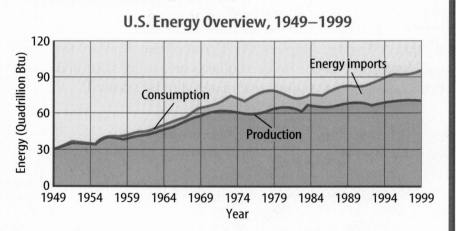

U.S. Energy Overview, 1949–1999

Identifying the Problem

The following graph shows the energy produced and consumed in the United States from 1949 to 1999. How does energy that is consumed by Americans compare with energy that is produced in the United States?

Solving the problem

1. Determine the approximate amount of energy produced in 1949 and in 1999 and how much it has increased in 50 years. Has it doubled or tripled?

2. Do the same for consumption. Has it doubled or tripled?

3. Using your answers for steps 1 and 2 and the graph, where does the additional energy that is needed come from? Give some examples.

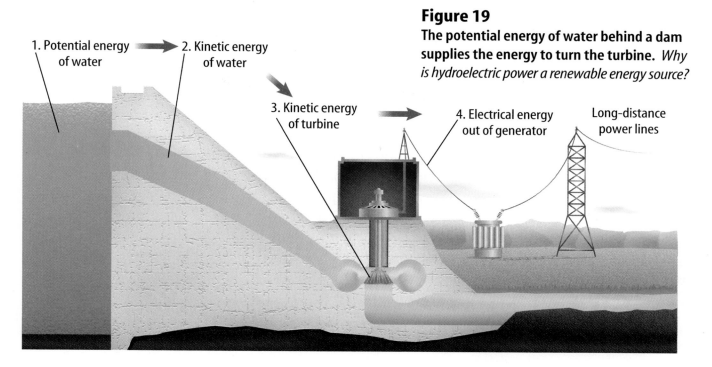

1. Potential energy of water → 2. Kinetic energy of water

3. Kinetic energy of turbine → 4. Electrical energy out of generator

Long-distance power lines

Figure 19
The potential energy of water behind a dam supplies the energy to turn the turbine. *Why is hydroelectric power a renewable energy source?*

Alternative Sources of Energy

Electrical energy can be generated in several ways. However, each has disadvantages that can affect the environment and the quality of life for humans. Research is being done to develop new sources of energy that are safer and cause less harm to the environment. These sources often are called **alternative resources.** These alternative resources include solar energy, wind, and geothermal energy.

Solar Energy

The Sun is the origin of almost all the energy that is used on Earth. Because the Sun will go on producing an enormous amount of energy for billions of years, the Sun is an inexhaustible source of energy. An **inexhaustible resource** is an energy source that can't be used up by humans.

Each day, on average, the amount of solar energy that strikes the United States is more than the total amount of energy used by the entire country in a year. However, less than 0.1 percent of the energy used in the United States comes directly from the Sun. One reason is that solar energy is more expensive to use than fossil fuels. However, as the supply of fossil fuels decreases, the cost of finding and mining these fuels might increase. Then, it may be cheaper to use solar energy or other energy sources to generate electricity and heat buildings than to use fossil fuels.

Reading Check *What is an inexhaustible energy source?*

Mini LAB

Building a Solar Collector

Procedure
1. Line a **large pot** with **black plastic** and fill with **water.**
2. Stretch **clear-plastic wrap** over the pot and tape it taut.
3. Make a slit in the top and slide a **thermometer** or a **computer probe** into the water.
4. Place your solar collector in direct sunlight and monitor the temperature change every 3 min for 15 min.
5. Repeat your experiment without using any black plastic.

Analysis
1. Graph the temperature changes in both setups.
2. Explain how your solar collector works.

Collecting the Sun's Energy Two types of collectors capture the Sun's rays. If you look around your neighborhood, you might see large, rectangular panels attached to the roofs of buildings or houses. If, as in **Figure 20A,** pipes come out of the panel, it is a thermal collector. Using a black surface, a thermal collector heats water by directly absorbing the Sun's radiant energy. Water circulating in this system can be heated to about 70°C. The hot water can be pumped through the house to provide heat. Also, the hot water can be used for washing and bathing. If the panel has no pipes, it is a photovoltaic (foh toh vol TAY ihk) collector, like the one pictured in **Figure 20B.** A **photovoltaic** is a device that transforms radiant energy directly into electrical energy. Photovoltaics are used to power calculators and satellites, including the *International Space Station.*

Reading Check *What does a photovoltaic do?*

Geothermal Energy

Imagine you could take a journey to the center of Earth—down to about 6,400 km below the surface. As you went deeper and deeper, you would find the temperature increasing. In fact, after going only about 3 km, the temperature could have increased enough to boil water. At a depth of 100 km, the temperature could be over 900°C. The heat generated inside Earth is called geothermal energy. Some of this heat is produced when unstable radioactive atoms inside Earth decay, converting nuclear energy to thermal energy.

At some places deep within Earth the temperature is hot enough to melt rock. This molten rock, or magma, can rise up close to the surface through cracks in the crust. During a volcanic eruption, magma reaches the surface. In other places, magma gets close to the surface and heats the rock around it.

Geothermal Reservoirs In some regions where magma is close to the surface, rainwater and water from melted snow can seep down to the hot rock through cracks and other openings in Earth's surface. The water then becomes hot and sometimes can form steam. The hot water and steam can be trapped under high pressure in cracks and pockets called geothermal reservoirs. In some places the hot water and steam are close enough to the surface to form hot springs and geysers.

Figure 20
Solar energy can be collected and utilized by individuals using **A** thermal collectors or **B** photovoltaic collectors.

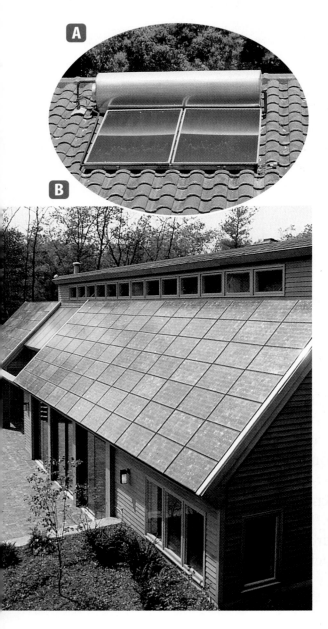

Geothermal Power Plants In places where the geothermal reservoirs are less than several kilometers deep, wells can be drilled to reach them. The hot water and steam produced by geothermal energy then can be used by geothermal power plants, like the one in **Figure 21,** to generate electricity.

Most geothermal reservoirs contain hot water under high pressure. **Figure 22** shows how these reservoirs can be used to generate electricity. While geothermal power is an inexhaustible source of energy, geothermal power plants can be built only in regions where geothermal reservoirs are close to the surface, such as in the western United States.

Heat Pumps Geothermal heat helps keep the temperature of the ground at a depth of several meters at a nearly constant temperature of about 10° to 20°C. This constant temperature can be used to cool and heat buildings by using a heat pump.

A heat pump contains a water-filled loop of pipe that is buried to a depth where the temperature is nearly constant. In summer the air is warmer than this underground temperature. Warm water from the building is pumped through the pipe down into the ground. The water cools and then is pumped back to the house where it absorbs more heat, and the cycle is repeated. During the winter, the air is cooler than the ground below. Then, cool water absorbs heat from the ground and releases it into the house.

Figure 21
This geothermal power plant in Nevada produces enough electricity to power about 50,000 homes.

Figure 22
The hot water in a geothermal reservoir is used to generate electricity in a geothermal power plant.

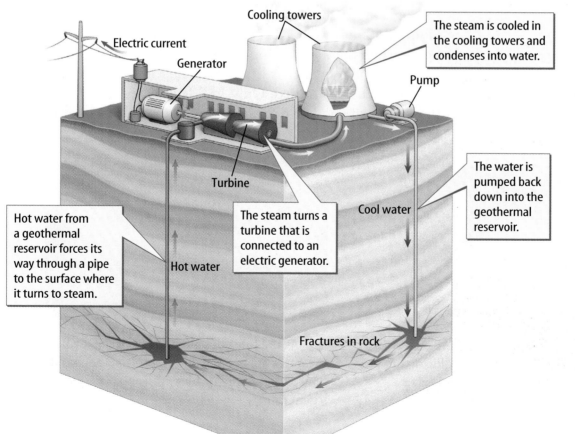

Electric current

Cooling towers

The steam is cooled in the cooling towers and condenses into water.

Generator

Pump

Turbine

The steam turns a turbine that is connected to an electric generator.

Cool water

The water is pumped back down into the geothermal reservoir.

Hot water from a geothermal reservoir forces its way through a pipe to the surface where it turns to steam.

Hot water

Fractures in rock

Energy from the Oceans

The ocean is in constant motion. If you've been to the seashore you've seen waves roll in. You may have seen the level of the ocean rise and fall over a period of about a half day. This rise and fall in the ocean level is called a tide. The constant movement of the ocean is an inexhaustible source of mechanical energy that can be converted into electric energy. While methods are still being developed to convert the motion in ocean waves to electric energy, several electric power plants using tidal motion have been built.

Figure 23
This tidal power plant in Annapolis Royal, Nova Scotia, is the only operating tidal power plant in North America.

Using Tidal Energy A high tide and a low tide each occur about twice a day. In most places the level of the ocean changes by less than a few meters. However, in some places the change is much greater. In the Bay of Fundy in Eastern Canada, the ocean level changes by 16 m between high tide and low tide. Almost 14 trillion kg of water move into or out of the bay between high and low tide.

Figure 23 shows an electric power plant that has been built along the Bay of Fundy. This power plant generates enough electric energy to power about 12,000 homes. The power plant is constructed so that as the tide rises, water flows through a turbine that causes a electric generator to spin, as shown in **Figure 24A**. The water is then trapped behind a dam. When the tide goes out, the trapped water behind the dam is released through the turbine to generate more electricity, as shown in **Figure 24B.** Each day electric power is generated for about ten hours when the tide is rising and falling.

While tidal energy is a nonpolluting, inexhaustible energy source, its use is limited. Only in a few places is the difference between high and low tide large enough to enable a large electric power plant to be built.

Figure 24
A tidal power plant can generate electricity when the tide is coming in and going out.

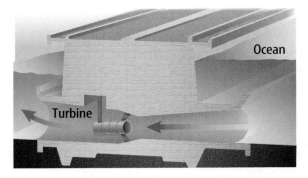

A As the tide comes in, it turns a turbine connected to a generator. When high tide occurs, gates are closed that trap water behind a dam.

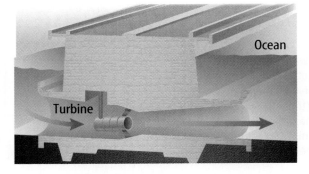

B As the tide goes out and the ocean level drops, the gates are opened and water from behind the dam flows through the turbine, causing it to spin and turn a generator.

Wind

Wind is another inexhaustible supply of energy. Modern windmills, like the ones in **Figure 25,** convert the kinetic energy of the wind to electrical energy. The propeller is connected to a generator so that electrical energy is generated when wind spins the propeller. These windmills produce almost no pollution. Some disadvantages are that windmills produce noise and that large areas of land are needed. Also, studies have shown that birds sometimes are killed by windmills.

Conserving Energy

Fossil fuels are a valuable resource. Not only are they burned to provide energy, but oil and coal also are used to make plastics and other materials. One way to make the supply of fossil fuels last longer is to use less energy. Reducing the use of energy is called conserving energy.

You can conserve energy and also save money by turning off lights and appliances such as televisions when you are not using them. Also keep doors and windows closed tightly when it's cold or hot to keep heat from leaking out of or into your house. Energy could also be conserved if buildings are properly insulated, especially around windows. The use of oil could be reduced if cars were used less and made more efficient, so they went farther on a liter of gas. Recycling materials such as aluminum cans and glass also helps conserve energy.

Figure 25
Windmills work on the same basic principles as a power plant. Instead of steam turning a turbine, wind turns the rotors. *What are some of the advantages and disadvantages of using windmills?*

Section 3 Assessment

1. What is the ultimate source of most of the energy stored on Earth?

2. What is a renewable resource? Give an example of a renewable and nonrenewable resource and explain the difference.

3. Explain why a heat pump is able to cool a building in the summer and heat the same building in the winter.

4. What are the disadvantages of using hydroelectricity and solar energy?

5. **Thinking Critically** Explain whether or not the following statement is true: All energy on Earth can be traced back to the Sun.

Skill Builder Activities

6. **Using an Electronic Spreadsheet** Use a spreadsheet to compare the effects on the environment of using fossil fuels, nuclear energy, and dams to produce electricity. Include in your spreadsheet the environmental effects of obtaining, transforming, and distributing the energy. **For more help, refer to the** Technology Skill Handbook.

7. **Using Proportions** As you go deeper into Earth, it becomes hotter. Using the information from this section, calculate the increase in temperature at a depth of 200 m below Earth's surface. **For more help, refer to the** Math Skill Handbook.

Activity *Use the Internet*

Energy to Power Your Life

Over the past 100 years, the amount of energy used in the United States and elsewhere has greatly increased. Today, a number of energy sources are available, such as coal, oil, natural gas, nuclear energy, hydroelectric power, wind, and solar energy. Some of these energy sources are being used up and are nonrenewable, but others are replaced as fast as they are used and, therefore, are renewable. Some energy sources are so vast that human usage has almost no effect on the amount available. These energy sources are inexhaustible.

Think about the types of energy you use at home and school every day. In this activity, you will investigate how and where energy is produced, and how it gets to you. You will also investigate alternative ways energy can be produced, and whether these sources are renewable, nonrenewable, or inexhaustible.

Recognize the Problem

What are the sources of the energy you use every day?

Form a Hypothesis

When you wake up in the morning and turn on a light, you use electrical energy. When you ride to school in a car or bus, its engine consumes chemical energy. What other types of energy do you use? Where is that energy produced? Which energy sources are nonrenewable, which are renewable, and which are inexhaustible? What are other sources of energy that you could use instead?

Local Energy Information	
Energy Type	
Where is that energy produced?	
How is that energy produced?	
How is that energy delivered to you?	
Is the energy source renewable, nonrenewable, or inexhaustible?	
What type of alternative energy source could you use instead?	

Goals
- **Identify** how energy you use is produced and delivered.
- **Investigate** alternative sources for the energy you use.
- **Outline** a plan for how these alternative sources of energy could be used.

Data Source
SCIENCEOnline Go to the Glencoe Science Web site at **science.glencoe.com** for more information about sources of energy and for data collected by other students.

Test Your Hypothesis

Plan

1. Think about the activities you do every day and the things you use. When you watch television, listen to the radio, ride in a car, use a hair drier, or turn on the air conditioning, you use energy. Select one activity or appliance that uses energy.

2. **Identify** the type of energy that is used.

3. **Investigate** how that energy is produced and delivered to you.

4. **Determine** if the energy source is renewable, nonrenewable, or inexhaustible.

5. If your energy source is nonrenewable, describe how the energy you use could be produced by renewable sources.

Do

1. Make sure your teacher approves your plan before you start.

2. Organize your findings in a data table, similar to the one that is shown.

3. Go to the Glencoe Science Web site at **science.glencoe.com** to post your data.

Analyze Your Data

1. **Describe** the process for producing and delivering the energy source you researched. How is it created, and how does it get to you?

2. How much of the energy you use every day comes from the energy source you investigated?

3. Is the energy source you researched renewable, nonrenewable, or inexhaustible? Why?

4. What other renewable or inexhaustible energy sources are used, or could be used, to generate electricity in your area?

Draw Conclusions

1. If the energy source you investigated is nonrenewable, describe how you could reduce your use of this energy source.

2. What alternative sources of energy could you use for everyday energy needs? On the computer, create a plan for using renewable or inexhaustible sources.

Communicating Your Data

SCIENCE Online Find this *Use the Internet* activity on the Glencoe Science Web site at **science.glencoe.com**. Post your data in the table that is provided. **Compare** your data to those of other students. **Combine** your data with those of other students and make inferences using the combined data.

Energy to Burn

Did you know...

...**Garbage—paper, vegetation, animal waste, and more**—could be a huge source of energy. Garbage converted to fuel can be used for heating, cooking, transportation, and electricity production. There are approximately 140 garbage-burning power plants in the United States. One truckload of garbage could produce energy equal to about 21 barrels of oil.

The energy released by the average hurricane is equal to about 200 times the total energy produced by all of the world's power plants. Almost all of this energy is released as heat when raindrops form.

...**The energy Earth gets each half hour from the Sun** is enough to meet the world's demands for a year. Renewable and inexhaustible resources, including the Sun, account for only 18 percent of the energy that is used worldwide.

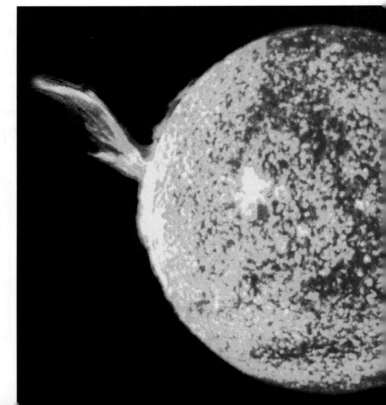

...Gas isn't the only source of energy for cars. Hybrid cars that use both gasoline engines and electric motors are being developed. These cars get higher gas mileage than cars powered by gasoline engines alone—over 60 miles per gallon.

...The largest energy failure in history occurred in 1965 and left 30 million people without electricity at the same time. The problem began because of a faulty relay switch in Ontario, Canada. An area from Ontario through most of New England was affected.

Global Energy Use 1970–2000

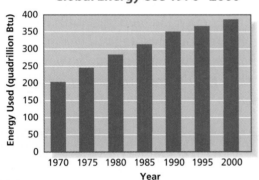

...The Calories in one medium apple will give you enough energy to walk for about 15 min, swim for about 10 min, or jog for about 9 min.

Do the Math

1. If 180 million people were living in the United States in 1965 and 24 million of them lost power, approximately what percentage of U.S. residents lost power?
2. How many liters of gas would you save by taking the hybrid car on a 174-km trip rather than taking the vehicle with the lowest gas mileage?
3. If walking for 15 min requires 80 Calories of fuel (from food), how many Calories would someone need to consume to walk for 1 h?

Go Further

Where would you place solar collectors in the United States? Why? For more information on solar energy, go to **science.glencoe.com**.

Reviewing Main Ideas

Section 1 What is energy?

1. Energy is the ability to cause change. Energy is found in many forms.

2. Moving objects have kinetic energy. The potential energy of an object depends on its height and mass.

3. Potential energy is the energy of position. Radiant energy is the energy of light.

4. Electric current carries electrical energy, and atomic nuclei contain nuclear energy. *What are all the forms of energy that are represented in this picture?*

Section 2 Energy Transformations

1. Energy can be transformed from one form to another. Energy transformations cause changes to occur.

2. All energy transformations obey the law of conservation of energy, which means no energy is ever created or destroyed.

3. Chemical and electrical energy can be converted into other forms of energy such as radiant energy and thermal energy. Thermal energy moves from warm to cool objects.

4. Power plants convert a source of energy into electrical energy. The kinetic energy of steam spins a turbine which causes a generator to spin. The spinning gernerator produces electricity.

Section 3 Sources of Energy

1. Fossil fuels and nuclear energy are nonrenewable energy sources. The use of each of these energy sources produces waste products.

2. Renewable and inexhaustible energy sources include hydroelectric, solar, geothermal, and wind energy. *Why is the energy source used by this hydroelectric plant a renewable source of energy?*

3. Energy shortages might be prevented by conserving energy. Each energy source has advantages and disadvantages. *Look at the two photographs below of the same view of New York but at different times. Explain the difference.*

FOLDABLES
Reading & Study Skills

After You Read

Write what you learned about the types, sources and transformation of energy under the Learned tab of your Know-Want-Learn Study Fold.

Visualizing Main Ideas

Use the following terms and phrases to complete the concept map about energy sources: fossil fuels, hydroelectric, solar, wind, oil, coal, photovoltaic, *and* nonrenewable resources.

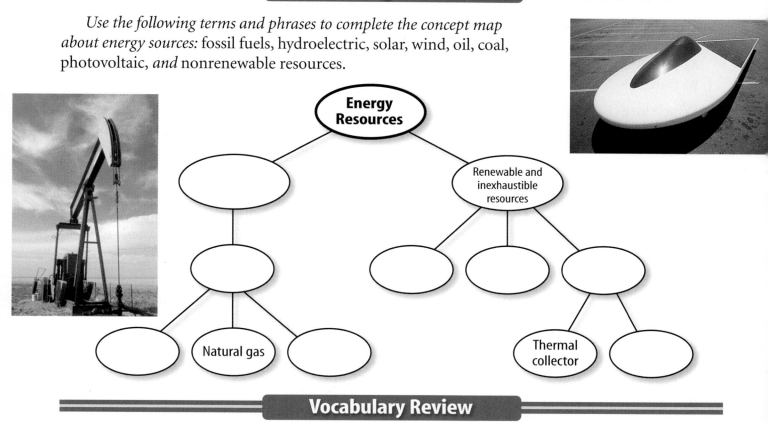

Vocabulary Review

Vocabulary Words

a. alternative resource
b. chemical energy
c. electrical energy
d. energy
e. generator
f. inexhaustible resource
g. kinetic energy
h. law of conservation of energy
i. nonrenewable resource
j. nuclear energy
k. photovoltaic
l. potential energy
m. radiant energy
n. renewable resource
o. thermal energy
p. turbine

Using Vocabulary

For each set of terms below, explain the relationship that exists.

1. electrical energy, nuclear energy
2. turbine, generator
3. photovoltaic, radiant energy, electrical energy
4. renewable resource, inexhaustible resource
5. potential energy, kinetic energy
6. kinetic energy, electrical energy, generator
7. thermal energy, radiant energy
8. law of conservation of energy, energy transformations
9. nonrenewable resource, chemical energy

THE PRINCETON REVIEW Study Tip

Practice reading graphs and charts. Make a table that contains the same information as a graph does.

Checking Concepts

1. Objects that are able to fall have what type of energy?
   A) kinetic   C) potential
   B) radiant   D) electrical

2. Which form of energy does light have?
   A) electrical   C) kinetic
   B) nuclear   D) radiant

3. Muscles perform what type of energy transformation?
   A) kinetic to potential
   B) kinetic to electrical
   C) thermal to radiant
   D) chemical to kinetic

4. Photovoltaics perform what type of energy transformation?
   A) thermal to radiant
   B) kinetic to electrical
   C) radiant to electrical
   D) electrical to thermal

5. Which form of energy does food have?
   A) chemical   C) radiant
   B) potential   D) electrical

6. Solar energy, wind, and geothermal are what type of energy resource?
   A) inexhaustible   C) nonrenewable
   B) inexpensive   D) chemical

7. Which of the following is a nonrenewable source of energy?
   A) hydroelectricity C) wind
   B) nuclear   D) solar

8. Which of the following does NOT require a generator to generate electricity?
   A) solar   C) hydroelectric
   B) wind   D) nuclear

9. Which of the following are fossil fuels?
   A) gas   C) oil
   B) coal   D) all of these

10. From where does the surface of Earth acquire most of its energy?
    A) radioactivity   C) chemicals
    B) Sun   D) wind

Thinking Critically

11. Explain how the motion of a swing illustrates the transformation between potential and kinetic energy.

12. A skateboard that is coasting along a flat surface will slow down and come to a stop. Explain what happens to the kinetic energy of the skateboard.

13. Describe the energy transformations that occur in the process of toasting a bagel in an electric toaster.

14. In what ways is the formation of coal like the formation of oil and natural gas? How is it different?

15. Explain the difference between the law of conservation of energy and conserving energy. How can conserving energy help prevent energy shortages?

Developing Skills

16. **Researching Information** Find out how spacecraft, such as *Galileo*, obtain the energy they need to operate as they travel through the solar system.

17. Concept Mapping Complete this concept map about energy.

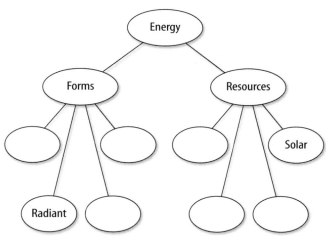

18. Classifying A proposal has been made to use wheat as a source of biomass energy. It can be made into alcohol, which can be burned in engines. How would you classify this source of energy? What are some advantages and disadvantages in using this as a source of energy? Explain.

Performance Assessment

19. Multimedia Presentation Alternative sources of energy that weren't discussed include biomass energy, wave energy, and hydrogen fuel cells. Research an alternative energy source and then prepare a digital slide show about the information you found. Use the concepts you learned from this chapter to inform your classmates about the future prospects of using such an energy source on a large scale.

TECHNOLOGY

Go to the Glencoe Science Web site at **science.glencoe.com** or use the **Glencoe Science CD-ROM** for additional chapter assessment.

THE PRINCETON REVIEW **Test Practice**

Throughout the course of one day, you engage in dozens of energy transformations. The table below gives some examples of different energy transformations.

Types of Energy Transformation	
Energy Transformation	**Example**
Potential → Kinetic	Ball rolling down a hill
Kinetic → Potential	A pebble tossed upward
Electrical → Radiant	A desk lamp
Chemical → Thermal	Burning fossil fuels
	Music from a radio

1. Which energy transformation occurs when coal is burned in a stove?
   A) potential → kinetic
   B) kinetic → potential
   C) electrical → radiant
   D) chemical → thermal

2. Which of these energy transformations will complete the table?
   F) electrical → radiant
   G) sound → electrical
   H) electrical → sound
   J) electrical → chemical

3. An image displayed on a computer screen most closely matches which example in terms of energy transformation?
   A) ball rolling down a hill
   B) a desk lamp
   C) burning fossil fuels
   D) music from a radio

Thermal Energy

On a sunny day you can feel the energy from the Sun as heat. The dishes shown here collect some of that energy and convert it into other forms of energy. How does heat energy travel through space? And how can heat energy be used to make cars move and refrigerate food? In this chapter you will learn about thermal energy and heat. You also will learn how heat is transferred from place to place and how the flow of heat can be controlled using different materials.

What do you think?

Science Journal Can you guess what this is a picture of? Here's a hint: *The one shown here is old, but you use a modern version of it every day.* In your Science Journal, write what you think it is and how you think it worked.

When you leave a glass of ice water on a kitchen table, the ice gradually melts and the temperature of the water increases. No matter how many times you leave a glass of ice water in a warm room, the water always gets warmer, never colder. What is temperature, and why does the temperature of the water always increase, but never decrease? In this activity you will explore one way of determining temperature.

Test Your Built-in Thermometer

1. Obtain three pans. Fill one pan with lukewarm water. Fill a second pan with cold water and crushed ice. Fill a third pan with very warm tap water. Label each pan.

2. Soak one of your hands in the warm water for a few minutes. Remove your hand from the warm water and put it in the lukewarm water. Does the lukewarm water feel cool or warm?

3. Now soak your hand in the cold water for a few minutes. Remove your hand from the cold water and place it in the lukewarm water. Does the lukewarm water feel cool or warm?

Observe

Write a paragraph in your Science Journal discussing whether your sense of touch would make a useful thermometer.

Before You Read

FOLDABLES
Reading & Study Skills

Making a Main Ideas Study Fold Make the following Foldable to help you identify the major topics about thermal energy.

1. Place a sheet of paper in front of you so the short side is at the top. Fold the top third of the paper down and the bottom third up.

2. Open the paper and label the three sections *Temperature, Thermal Energy,* and *Heat.*

3. Before you read the chapter, write what you know about each on your Foldable. As you read the chapter, add to and correct what you have written.

Temperature and Thermal Energy

What You'll Learn
- **Explain** how temperature is related to kinetic energy.
- **Describe** three scales used for measuring temperature.
- **Define** thermal energy.

Vocabulary
temperature
thermal energy

Why It's Important
Temperature and thermal energy influence the movement of heat in the environment.

What's Hot? What's Cold?

Imagine it's a hot day and you jump into a swimming pool to cool off. When you first hit the water, you might think it feels cold. Perhaps someone else, who has been swimming for a few minutes, thinks the water feels warm. When you swim in water, touch a hot pan, or swallow a cold drink, your sense of touch tells you whether something is hot or cold. However, the words *cold*, *warm*, and *hot* can mean different things to different people.

Temperature How hot or cold something feels is related to its temperature. To understand temperature, think of a glass of water sitting on a table. The water might seem perfectly still, but is it? **Figure 1** shows that water is made of molecules that are in constant motion. Because these molecules are always moving, they have energy of motion, or kinetic energy.

However, all water molecules don't move at the same speed. Some are moving faster and some are moving slower. **Temperature** is a measure of the average value of the kinetic energy of the molecules in a substance. The more kinetic energy the molecules have, the higher the temperature. Molecules have more kinetic energy when they are moving faster. So the higher the temperature, the faster the molecules are moving, as shown in **Figure 1**.

Figure 1
The temperature of a substance depends on how fast its molecules are moving. Water molecules are moving faster in the hot water on the left than in the cold water on the right.

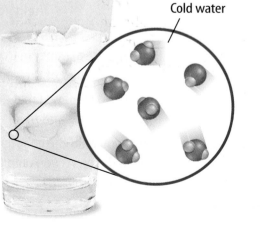

Cold water

Hot water

Thermal Expansion It wasn't an earthquake that caused the sidewalk to buckle in **Figure 2.** Hot weather caused the concrete to expand so much that it cracked, and the pieces squeezed each other upward. When the temperature of an object is increased, its molecules speed up and tend to move farther apart. This causes the object to expand. When the object is cooled, its molecules slow down and move closer together. This causes the object to shrink, or contract.

Almost all substances expand when they are heated and contract when they are cooled. The amount of expansion or contraction depends on the type of material and the change in temperature. For example, liquids usually expand more than solids. Also, the greater the change in temperature, the more an object expands or contracts.

> ☑ **Reading Check** *Why do materials expand when their temperatures increase?*

Measuring Temperature

The temperature of an object depends on the average kinetic energy of all the molecules in an object. However, molecules are so small and objects contain so many of them, that it is impossible to measure the kinetic energy of all the individual molecules.

Instead, a practical way to measure temperature is to use a thermometer. Thermometers usually use the expansion and contraction of materials to measure temperature. One common type of thermometer uses a glass tube containing a liquid. When the temperature of the liquid increases, it expands so that the height of the liquid in the tube depends on the temperature.

Temperature Scales To be able to give a number for the temperature, a thermometer has to have a temperature scale. Two common temperature scales are the Fahrenheit and Celsius scales, shown in **Figure 3.**

On the Fahrenheit scale, the freezing point of water is given the temperature 32°F and the boiling point 212°F. The space between the boiling point and the freezing point is divided into 180 equal degrees. The Fahrenheit scale is used mainly in the United States.

On the Celsius temperature scale, the freezing point of water is given the temperature 0°C and the boiling point is given the temperature 100°C. Because there are only 100 Celsius degrees between the boiling and freezing point of water, Celsius degrees are bigger than Fahrenheit degrees. Which is warmer, 50°F or 50°C?

Figure 2
Most objects expand as their temperatures increase. Pieces of this concrete sidewalk forced each other upward when the concrete expanded on a hot day.

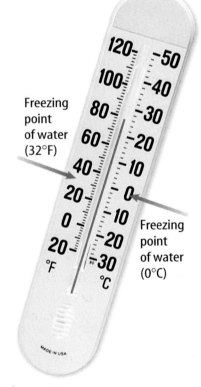

Freezing point of water (32°F)

Freezing point of water (0°C)

Figure 3
The Fahrenheit and Celsius scales are the most commonly used temperature scales. *Which has the most degrees between the boiling and freezing points of water?*

Converting Fahrenheit and Celsius Temperatures on the Fahrenheit scale can be converted to Celsius temperatures using this formula.

$$°C = \left(\frac{5}{9}\right)(°F - 32)$$

For example, to convert 68°F to Celsius, first subtract 32, multiply by 5, then divide by 9. The result is 20°C.

The Kelvin Scale Another temperature scale that is sometimes used is the Kelvin Scale. On this scale, 0 K is the lowest temperature an object can have. This temperature is known as absolute zero. The size of a degree on the Kelvin scale is the same as on the Celsius scale. You can change from Kelvin degrees to Celsius degrees by subtracting 273 from the Kelvin temperature.

$$°C = K - 273$$

Math Skills Activity

Converting Fahrenheit to Celsius Temperatures

Example Problem

On a hot summer day, a Fahrenheit thermometer shows the temperature to be 86°F. What is this temperature on the Celsius scale?

Solution

1 *This is what you know:*   Fahrenheit temperature: °F = 86

2 *This is what you need to know:*   Celsius temperature: °C

3 *This is the equation you need to use:* $°C = \left(\frac{5}{9}\right)(°F - 32)$

4 *Substitute the known values:* $°C = \left(\frac{5}{9}\right)(86 - 32)$

$$= \left(\frac{5}{9}\right)(54) = 30$$

Check your answer by multiplying it by $\frac{9}{5}$ and adding 32.

Do you calculate the same temperature that was given?

Practice Problem

A person's body temperature is 98.6°F. What is this temperature on the Celsius scale?

For more help, refer to the Math Skill Handbook.

Thermal Energy

The temperature of an object is related to the average kinetic energy of its molecules. But molecules also have potential energy. Potential energy is energy that the molecules have that can be converted into kinetic energy. The sum of the kinetic and potential energy of all the molecules in an object is the **thermal energy** of the object.

The Potential Energy of Molecules When you hold a ball above the ground, it has potential energy. When you drop the ball, its potential energy is converted into kinetic energy as the ball falls toward Earth. It is the attractive force of gravity between Earth and the ball that gives the ball potential energy.

The molecules in a material also exert attractive forces on each other. As a result, the molecules in a material have potential energy. As the molecules get closer together or farther apart, their potential energy changes.

Increasing Thermal Energy Temperature and thermal energy are different. Suppose you have two glasses with the same amount of milk, and at the same temperature. If you pour both glasses of milk into a pitcher, as shown in **Figure 4,** the temperature of the milk won't change. However, because there are more molecules of milk in the pitcher, the thermal energy of the milk in the pitcher is greater. Because the amount of milk doubled, the thermal energy of the milk doubled.

Figure 4
When two substances of equal temperature are combined, the temperature remains the same, but the thermal energy increases.

Section ① Assessment

1. Explain the difference between temperature and thermal energy.
2. Write a formula for converting from Fahrenheit to Kelvin.
3. How are temperature and kinetic energy related?
4. How does a thermometer use the thermal expansion of a material to measure temperature?
5. **Think Critically** You have two identical bottles of soda. One is placed in the Sun, the other in an ice chest. Which has more thermal energy? Explain.

Skill Builder Activities

6. **Making and Using Tables** Make a table showing the Fahrenheit, Celsius, and Kelvin temperature of the following: *normal body temperature, air temperature on a summer day,* and *air temperature on a winter day.* **For more help, refer to the** Science Skill Handbook.

7. **Solving One-Step Equations** The turkey you're cooking for dinner will be ready when it reaches an internal temperature of 180°F. Convert this temperature to °C and K. **For more help, refer to the** Math Skill Handbook.

Heat

As You Read

What You'll Learn

- **Explain** the difference between thermal energy and heat.
- **Describe** three ways heat is transferred.
- **Identify** materials that are insulators or conductors.

Vocabulary

heat   conductor
conduction   specific heat
radiation   thermal pollution
convection

Why It's Important

A knowledge of heat and how it is transferred will help you learn to use energy more efficiently.

Heat and Thermal Energy

It's the heat of the day. Heat the oven to 375°F. A heat wave has hit the Midwest. You've often heard the word *heat,* but what is it? Is it something you can see? Can an object have heat? Is heat anything like thermal energy? **Heat** is thermal energy that is transferred from one object to another when the objects are at different temperatures. The amount of heat that is transferred when two objects are brought into contact depends on the difference in temperature between the objects.

For example, no heat is transferred when two pots of boiling water are touching, because the water in both pots is at the same temperature. However, heat is transferred from the pot of hot water in **Figure 5** that is touching a pot of cold water. The hot water cools down and the cold water gets hotter. Heat continues to be transferred until both objects have the same temperature.

Transfer of Heat When heat is transferred, thermal energy always moves from warmer to cooler objects. Heat never flows from a cooler object to a warmer object. The warmer object loses thermal energy and becomes cooler as the cooler object gains thermal energy and becomes warmer. This process of heat transfer can occur in three ways—by conduction, radiation, or convection.

Figure 5
Heat is transferred only when two objects are at different temperatures. Heat always moves from the warmer object to the cooler object.

Conduction When you eat hot pizza, you experience conduction. As the hot pizza touches your mouth, heat moves from the pizza to your mouth. This transfer of heat by direct contact is called **conduction.** Conduction occurs when the particles of one substance collide with the particles of another substance and transfer some kinetic energy.

Imagine holding an ice cube in your hand, as in **Figure 6.** The faster-moving molecules in your warm hand bump against the slower-moving molecules in the cold ice. In these collisions, energy is passed from molecule to molecule. Heat flows from your warmer hand to the colder ice, and the slow-moving molecules in the ice move faster. As a result, the ice becomes warmer and its temperature increases. Molecules in your hand move more slowly as they lose thermal energy, and your hand becomes cooler.

Conduction occurs most easily in solids, where the bonds between atoms and molecules keep them close together. Because they are so close together, atoms and molecules in a solid need to move only a short distance before they bump into one another and transfer energy.

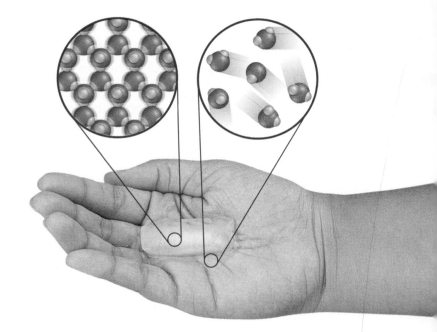

Figure 6
An ice cube in your hand melts because of conduction. The solid ice melts, becoming liquid water. Molecules in the water move faster than molecules in the ice. *What is being transferred as the ice cube melts?*

Reading Check *Why does conduction occur easily in solids?*

Radiation On a beautiful, clear day, you walk outside and notice the warmth of the Sun. You know that the Sun heats Earth, but how does this transfer of thermal energy occur? The heat transfer does not occur by conduction, because almost no matter exists between the Sun and Earth. Instead, heat is transferred from the Sun to Earth by radiation. Heat transfer by **radiation** occurs when energy is transferred by electromagnetic waves. These invisible waves carry energy through empty space, as well as through matter. The transfer of thermal energy by radiation can occur in empty space, as well as in solids, liquids, and gases.

The Sun is not the only source of radiation. All objects emit electromagnetic radiation, although warm objects emit more radiation than cool objects. The warmth you feel when you sit next to a fireplace is due to heat transferred by radiation from the fire to your skin.

Convection When you heat a pot of water on a stove, heat can be transferred through the water by another process besides conduction and radiation. In a gas or liquid, molecules can move much more easily then they can in a solid. As a result, the more energetic molecules can travel from one place to another, and carry their energy along with them. This transfer of thermal energy by the movement of molecules from one part of a material to another is called **convection.**

Transferring Heat by Convection As a pot of water is heated, heat is transferred by convection. First, thermal energy is transferred to the water molecules at the bottom of the pot from the stove. These water molecules move faster as their thermal energy increases. The faster-moving molecules tend to be farther apart than the slower-moving molecules in the cooler water above. Because the molecules are farther apart in the warm water, this water is less dense than the cooler water. As a result, the warm water rises and is replaced at the bottom of the pot by cooler water. The cooler water is heated, rises, and the cycle is repeated until all the water in the pan is at the same temperature.

Natural Convection Natural convection occurs when a warmer, less dense fluid is pushed away by a cooler, denser fluid. For example, imagine the shore of a lake. During the day, the water is cooler than the land. As shown in **Figure 7,** air above the warm land is heated by conduction. When the air gets hotter, its particles move faster and get farther from each other, making the air less dense. The cooler, denser air from over the lake flows in over the land, pushing the less dense air upward. You feel this movement of incoming cool air as wind. The cooler air then is heated by the land and also begins to rise.

Figure 7
Wind movement near a lake or ocean results from natural convection. Air is heated by the land and becomes less dense. Denser cool air rushes in, pushing the warm air up. The cooler air then is heated by the land and rises and the cycle is repeated.

Warm air

Cool air

Warm air

Cool air

Forced Convection Sometimes convection can be forced. Forced convection occurs when an outside force pushes a fluid, such as air or water, to make it move and transfer heat. A fan is one type of device that is used to move air. For example, computers use fans to keep their electronic components from getting too hot, which can damage them. The fan blows cool air onto the hot electronic components, as shown in **Figure 8.** Heat from the electronic components is transferred to the air around them by conduction. The warm air is pushed away as cool air rushes in. The hot components then continue to lose heat as the fan blows cool air over them.

Thermal Conductors

Why are cooking pans usually made of metal? Why does the handle of a metal spoon in a bowl of hot soup become warm? The answer to both questions is that metal is a good conductor. A **conductor** is any material that easily transfers heat. Some materials are good conductors because of the types of atoms or chemical compounds they are made up of.

☑ Reading Check *What is a conductor?*

Remember that an atom has a nucleus surrounded by one or more electrons. Certain materials, such as metals, have some electrons that are not held tightly by the nucleus and are freer to move around. These loosely held electrons can bump into other atoms and help transfer thermal energy. The best conductors of heat are metals such as gold and copper.

Mini LAB

Observing Convection

Procedure
1. Fill a **250-mL beaker** with room-temperature **water** and let it stand undisturbed for at least 1 min.
2. Using a **hot plate,** heat a small amount of water in a **50-mL beaker** until it is almost boiling.
   WARNING: *Do not touch the heated hot plate.*
3. Carefully drop a **penny** into the hot water and let it stand for about 1 min.
4. Take the penny out of the hot water with **metal tongs** and place it on a table. Immediately place the 250-mL beaker on the penny.
5. Using a **dropper,** gently place one drop of **food coloring** on the bottom of the 250-mL beaker of water.
6. Observe what happens in the beaker for several minutes.

Analysis
1. What happened when you placed the food coloring in the 250-mL beaker? Why?

Life Science
INTEGRATION

To survive in its arctic environment, a polar bear needs good insulation against the cold. Underneath its fur, a polar bear has 10 cm of insulating blubber. Research how blubber helps insulate the polar bear from the cold air and write your findings in your Science Journal.

Figure 9
The insulation in houses and buildings helps stop the transfer of heat between the air inside and air outside.

Thermal Insulators

If you're cooking food, you want the pan to conduct heat easily from the stove to your food, but you do not want the heat to move easily to the handle of the pan. An insulator is a material in which heat doesn't flow easily. Most pans have handles that are made from insulators. Liquids and gases are usually better insulators than solids are. Air is a good insulator, and many insulating materials contain air spaces that reduce the transfer of heat by conduction within the material. Materials that are good conductors, such as metals, are poor insulators, and poor conductors are good insulators.

Houses and buildings are made with insulating materials to reduce heat conduction between the inside and outside. Fluffy insulation like that shown in **Figure 9** is put in the walls. Some windows have double layers of glass that sandwich a layer of air or other insulating gas. This reduces the outward flow of heat in the winter and the inward flow of heat in the summer.

Heat Absorption

On a hot day, you can walk barefoot across the lawn, but the asphalt pavement of a street is too hot to walk on. Why is the pavement hotter than the grass? The change in temperature of an object as it absorbs heat depends on the material it is made of.

Specific Heat The amount of heat needed to change the temperature of a substance is related to its specific heat. The **specific heat** of a substance is the amount of heat needed to raise the temperature of 1 kg of that substance by 1°C.

More heat is needed to change the temperature of a material with a high specific heat than one with a low specific heat. For example, the sand on a beach has a lower specific heat than water has. When you're at the beach during the day, the sand feels much warmer than the water does. Radiation from the Sun warms the sand and the water. Because of its lower specific heat, the sand heats up faster than the water. At night, however, the sand feels cool and the water feels warmer. The temperature of the water changes more slowly than the temperature of the sand as they both lose thermal energy to the cooler night air.

Thermal Pollution

 **Life Science INTEGRATION**

Some electric power plants and factories that use water for cooling produce hot water as a by-product. If this hot water is released into an ocean, lake, or river, it will raise the temperature of the water nearby. This increase in the temperature of a body of water caused by adding warmer water is called **thermal pollution.** Rainwater that is heated after it falls on warm roads or parking lots also can cause thermal pollution if it runs off into a river or lake.

Effects of Thermal Pollution Increasing the water temperature causes fish and other aquatic organisms to use more oxygen. Because warmer water contains less dissolved oxygen than cooler water, some organisms can die due to a lack of oxygen. Also, in warmer water, many organisms become more sensitive to chemical pollutants, parasites, and diseases.

Reducing Thermal Pollution Thermal pollution can be reduced by cooling the warm water produced by factories, power plants, and runoff before it is released into a body of water. Cooling towers like the ones shown in **Figure 10** are used to cool the water from some power plants and factories. In some places the warm water is held in cooling ponds where it cools before it is released.

Figure 10
This power plant uses cooling towers to cool its waste water before releasing into the lake.

Section ② Assessment

1. Why isn't it correct to say that an object has heat?
2. Describe the three ways that heat can be transferred.
3. Look around your classroom and name some objects that are good insulators and some that are good conductors.
4. In the spring, the temperature of a lake increases more slowly than the temperature of the surrounding land. Explain.
5. **Think Critically** Is it better to have heating vents in your home near the floor or near the ceiling? Why?

Skill Builder Activities

6. **Recognizing Cause and Effect** England and southern Canada are at about the same latitude, and yet they have different climates. Canada usually has cold winters, but England usually has cool winters because of a nearby warm ocean current. Use what you've learned about heat transfer to explain this effect. **For more help, refer to the** Science Skill Handbook.
7. **Communicating** In your Science Journal, describe several examples of heat transfer by conduction in your everyday life. **For more help, refer to the** Science Skill Handbook.

Activity

Heating Up and Cooling Down

Do you remember how long it took for a cup of hot chocolate to cool before you could take a sip? In this activity, investigate how quickly liquids at different temperatures will heat up and cool down.

What You'll Investigate
How does the temperature of a liquid affect how quickly it warms or cools?

Materials
thermometers (5)
400-mL beakers (5)
stopwatch
watch with second hand
hotplate
Alternate materials

Goals
■ Measure the temperature change of water at different temperatures.
■ Infer how the rate of heating or cooling depends on the initial water temperature.

Safety Precautions

Do not use mercury thermometers. Use caution when heating with a hot plate. Hot and cold glass appears the same.

Procedure
1. Make a data table to record the temperature of water in five beakers every minute from 0 to 10 min.
2. Fill one beaker with 100 mL of water. Place the beaker on a hotplate and bring the water to a boil. *Safely* remove the hot beaker from the hotplate.

3. Record the water temperature in your data table at minute 0, and then every minute for 10 min.
4. Repeat step 3 starting with hot tap water, cold tap water, refrigerated water, and ice water with the ice removed.

Conclude and Apply
1. **Graph** your data. Plot the lines for all five beakers on one graph. **Label** the lines of your graph.
2. **Calculate** the rate of heating or cooling for the water in each beaker by subtracting the initial temperature of the water from the final temperature and then dividing this answer by 10 min.
3. **Infer** from your results how the difference between room temperature and the initial temperature of the water affected the rate at which it heated up or cooled down.

*C*ommunicating
Your Data
Share your data and graphs with other classmates and explain any differences among your data.

Engines and Refrigerators

Heat Engines

Cars have engines. Motorcycles have engines. Lawn mowers have engines. Engines are used everywhere. How do they work? An **engine** is any device that converts thermal energy into mechanical energy. One type of engine burns fuel to produce thermal energy. In an external combustion engine, such as the steam engine shown in **Figure 11,** the fuel is burned outside the engine. The burning fuel converts water into steam that pushes a piston. The moving piston can then do useful work.

✔ **Reading Check** *What is an engine?*

As You Read

What You'll Learn
- **Identify** what an engine does.
- **Describe** how an internal combustion engine works.
- **Explain** how refrigerators and air conditioners create cool environments.

Vocabulary
engine
internal combustion engine

Why It's Important
Engines help you travel every day, and refrigerators keep your food fresh and cold.

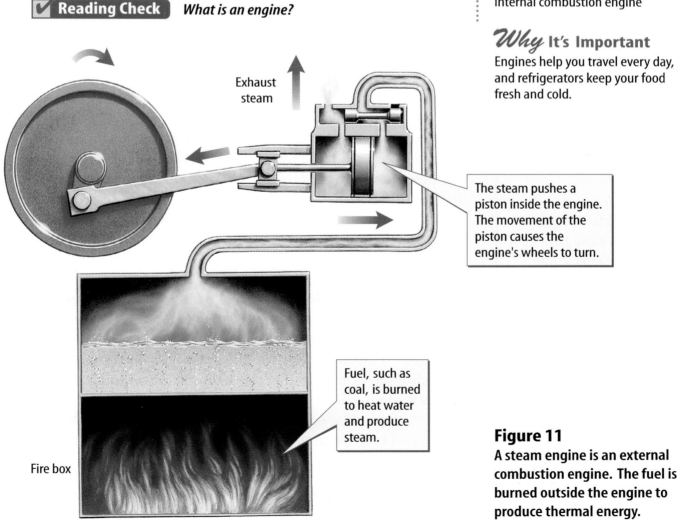

Exhaust steam

The steam pushes a piston inside the engine. The movement of the piston causes the engine's wheels to turn.

Fuel, such as coal, is burned to heat water and produce steam.

Fire box

Figure 11
A steam engine is an external combustion engine. The fuel is burned outside the engine to produce thermal energy.

Figure 12
Internal combustion engines are found in many tools and machines.

SCIENCE *Online*

Research Visit the Glencoe Science Web site at **science.glencoe.com** for more information about advancements in the design of internal combustion engines. Evaluate the advantages and disadvantages of these new designs with your classmates.

Internal Combustion Engines The type of engine you are probably most familiar with is the internal combustion engine. In **internal combustion engines,** the fuel burns in a combustion chamber inside the engine. Many machines, including cars, airplanes, buses, boats, trucks, and lawn mowers, use internal combustion engines, as shown in **Figure 12.**

Most cars have an engine with four or more combustion chambers, or cylinders. Usually the more cylinders an engine has, the more power it can produce. Each cylinder contains a piston that can move up and down. A mixture of fuel and air is injected into a combustion chamber and ignited by a spark. When the fuel mixture is ignited, it burns explosively and pushes the piston down. The up-and-down motion of the pistons turns a rod called a crankshaft, which turns the wheels of the car. **Figure 13** shows how an internal combustion engine converts thermal energy to mechanical energy in a process called the four-stroke cycle.

Several kinds of internal combustion engines have been designed. In diesel engines, the air in the cylinder is compressed to such a high pressure that the highly flammable fuel ignites without the need for a spark plug. Many lawn mowers use a two-stroke gasoline engine. The first stroke is a combination of intake and compression. The second stroke is a combination of power and exhaust.

✔ **Reading Check** *How does the burning of fuel mixture cause a piston to move?*

Figure 13

Most modern cars are powered by fuel-injected internal combustion engines that have a four-stroke combustion cycle. Inside the engine, thermal energy is converted into mechanical energy as gasoline is burned under pressure inside chambers known as cylinders. The steps in the four-stroke cycle are shown here.

EXHAUST STROKE

Exhaust valve

Exhaust gases

COMPRESSION STROKE

Fuel-air mixture

POWER STROKE

Spark plug

INTAKE STROKE

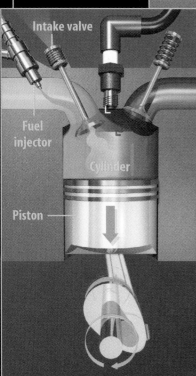

Intake valve

Fuel injector

Cylinder

Piston

Crankshaft

B The piston moves up, compressing the fuel-air mixture.

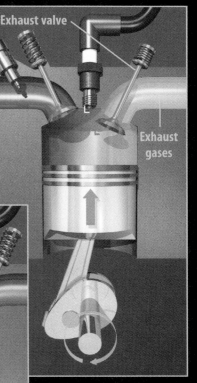

D The exhaust valve opens as the piston moves up, pushing the exhaust gases out of the cylinder.

C At the top of the compression stroke, a spark ignites the fuel-air mixture. The hot gases that are produced expand, pushing the piston down and turning the crankshaft.

A During the intake stroke, the piston inside the cylinder moves downward. As it does, air fills the cylinder through the intake valve, and a mist of fuel is injected into the cylinder.

Refrigerators

If thermal energy will only flow from something that is warm to something that is cool, how can a refrigerator be cooler inside than the air in the kitchen? A refrigerator is a heat mover. It absorbs heat from the food and other materials inside the refrigerator. Then it carries the heat to outside the refrigerator, where it is transferred to the surrounding air.

A refrigerator contains a material called a coolant that is pumped through pipes inside and outside the refrigerator. The coolant is the substance that carries heat from the inside to the outside of the refrigerator.

Absorbing Heat **Figure 14** shows how a refrigerator operates. Liquid coolant is forced up a pipe toward the freezer unit. The liquid passes through an expansion valve where it changes into a gas. When it changes into a gas, it becomes cold. The cold gas passes through pipes around the inside of the refrigerator. Because the coolant gas is so cold, it absorbs heat from inside the refrigerator, and becomes warmer.

Releasing Heat However, the gas is still colder than the outside air. So, the heat absorbed by the coolant cannot be transferred to the air. The warm coolant gas then passes through a compressor that compresses the gas. When the gas is compressed, it becomes warmer than room temperature. The gas then flows through the condenser coils, where it transfers heat to the cooler air in the room. As the coolant gas cools, it changes into a liquid. The liquid is pumped through the expansion valve, changes into a gas, and the cycle is repeated.

Figure 14
As refrigerant moves through the coils inside a refrigerator, it absorbs heat and evaporates. The refrigerant is recycled when it cools and condenses in the outside coils and then is brought back to the inside coils. *Where do the transfers of heat in a refrigerator occur?*

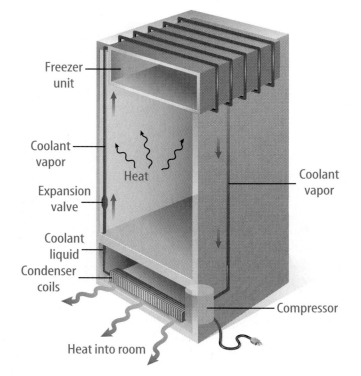

Freezer unit

Coolant vapor

Heat

Coolant vapor

Expansion valve

Coolant liquid

Condenser coils

Compressor

Heat into room

Air Conditioners Most air conditioners cool in the same way that a refrigerator does. You've probably seen air-conditioning units outside of many houses. As in a refrigerator, heat from inside the house is absorbed by the coolant within pipes inside the air conditioner. The coolant then is compressed by a compressor, and becomes warmer. The warmed coolant travels through pipes that are exposed to the outside air. Here the heat is transferred to the outside air.

Heat Pumps Some buildings use a heat pump for heating as well as cooling, as shown in **Figure 15.** Like an air conditioner or refrigerator, a heat pump moves heat from one place to another. When a heat pump is used for cooling, it removes thermal energy from the indoor air and transfers it outdoors. When it is used for heating, the heat pump absorbs thermal energy from the outdoor air or some other warm source and transfers this heat inside. The heat pump can reverse itself automatically. It can heat or cool depending on the outside temperature. In summer, the heat pump removes heat from the air inside the building and releases it outside. In winter, it removes heat from the outside ground or air and transfers it inside the house. In areas where the winter temperature is near or below zero, an additional heating coil sometimes is added to provide more heat.

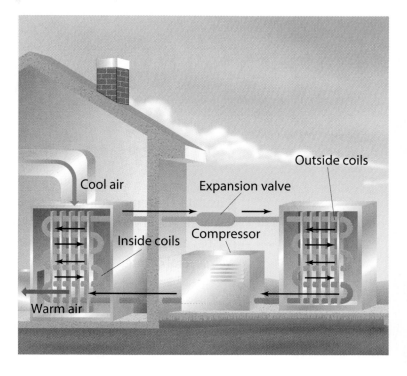

Figure 15
A heat pump can be used to heat and cool a building. In heating mode, the coolant absorbs heat through the outside coils. The coolant is warmed when it is compressed, and transfers heat to the room through the inside coils. In cooling mode, the coolant moves through the system in the opposite direction.

Section 3 Assessment

1. In an engine, thermal energy is converted into what form of energy?
2. What is the source of thermal energy in an internal combustion engine?
3. Why don't diesel engines use spark plugs?
4. Explain how a refrigerator keeps the food compartment cool.
5. **Think Critically** Why do you think a car has four or more cylinders rather than just one cylinder?

Skill Builder Activities

6. **Concept Mapping** Make an events-chain concept map showing the steps in a four-stroke cycle. **For more help, refer to the** Science Skill Handbook.

7. **Using Graphics Software** Using computer graphics or drawing software, make a diagram that shows a typical refrigeration cycle. **For more help, refer to the** Technology Skill Handbook.

Comparing Thermal Insulators

Insulated beverage containers are used to reduce heat transfer. What kinds of containers do you more commonly drink from? Aluminum soda cans? Paper, plastic, or foam cups? Glass containers? In this investigation, compare how well several different containers block heat transfer.

Recognize the Problem

Which types of beverage containers are most effective at blocking heat transfer from a hot drink?

Form a Hypothesis

Predict the temperature change of a hot liquid in several containers made of different materials over a time interval.

Goals

- **Predict** the temperature change of a hot drink in various types of containers over time.
- **Design** an experiment to test the hypothesis and collect data that can be graphed.
- **Interpret** the data.

Possible Materials

hotplate
large beaker
water
graduated cylinder
thermometers
various beverage containers (each about the same size and shape)
material to cover the containers
stopwatch
*watch with a second hand
tongs
thermal gloves or mitts
*Alternate materials

Safety Precautions 🥽 🧤

Use caution when heating liquids. Be sure to use tongs or thermal gloves when handling hot materials. Remember that hot and cold glass appears the same. Treat thermometers with care and keep them away from the edges of tables. Avoid using mercury thermometers.

Test Your Hypothesis

Plan

1. **Decide** what types of containers you will test. Design an experiment to test your hypothesis. This is a group activity, so make certain that everyone gets to contribute to the discussion.

2. **List** the materials you will use in your experiment. Describe exactly how you will use these materials. Which liquid will you test? What temperature will the liquid begin at? How will you cover the hot liquids in the container? What material will you use as a cover?

3. **Identify** the variables and controls in your experiment.

4. **Design** a data table in your Science Journal to record the observations you make.

Do

1. Ask your teacher to examine the steps of your experiment and your data table before you start.

2. To see the pattern of how well various containers retain heat, you will need to graph your data. What kind of graph will you use? Make certain you take enough measurements during the experiment to make your graph.

3. The time intervals between measurements should be the same. Be sure to keep track of time as the experiment goes along. For how long will you measure the temperature?

4. Carry out your investigation and record your observations.

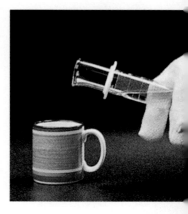

Analyze Your Data

1. **Graph** your data. Use one graph to show the data collected from all your containers. Label each line on your graph.

2. How can you tell by looking at your graphs which containers retain heat best?

3. Did the water temperature change as you had predicted? Use your data and graph to explain your answers.

Draw Conclusions

1. Why did the rate of temperature change vary among the containers? Did the size of the containers affect the rate of cooling?

2. **Conclude** which containers were the best insulators.

Communicating Your Data

Compare your data and graphs with other classmates and explain any differences in your results or conclusions.

SCIENCE AND
Society

SCIENCE
ISSUES
THAT AFFECT
YOU!

The Heat Is

You may live far from water, but still live on an island—a heat island

Here's a riddle: Rebecca and Julie were crossing a parking lot on a hot summer day. Rebecca had shoes on. Julie did not. They had gone only a short way when one of the girls broke into a run. Who was it and why did she run?

Dark materials, such as asphalt, absorb more heat than light materials. In extreme heat, it's even possible to fry an egg on dark pavement!

If you guess barefoot Julie, you're right. The hot asphalt of the parking lot scorched her feet, and Julie took off like a shot. Paving and building materials such as asphalt and concrete absorb more energy from the Sun and get hotter. Think about all the things that are made of asphalt and concrete in a city.

As far as the eye can see, there are buildings and parking lots, sidewalks and streets. The combined effect of these paved surfaces and towering structures can make a city sizzle in the summer. There's even a name for this effect. It's called the heat island effect.

Hot Times

You can think of a city as an island surrounded by an ocean of green trees and other vegetation. In the midst of those green trees, the air can be up to 8°C cooler than it is downtown. During the day in rural areas, the Sun's energy is absorbed by plants and soil. Some of this energy causes water to evaporate, so less energy is available to heat the surroundings. This keeps the temperature lower.

In cities, where there are fewer trees and plants, the buildings, streets, and sidewalks absorb most of the Sun's energy. And as more energy is absorbed, the temperature increases. As the temperature of the streets and buildings rises, they lose heat to cooler objects in their surroundings.

The temperature stops rising when heat energy is released at the same rate that energy from the Sun is absorbed.

Higher temperatures aren't the only problems caused by heat islands. People crank up their air conditioners for relief, so the use of energy skyrockets. Also, the added heat speeds up the rates of chemical reactions in the atmosphere. Smog is due to chemical reactions caused by the interaction of sunlight and vehicle emissions. So hotter air means more smog. And more smog means more health problems.

Cool Cures

Several U.S. cities are working with NASA scientists to come up with a cure for the summertime blues. For instance, dark materials absorb heat more efficiently than light materials. So painting buildings, especially roofs, white can reduce heat and save on cooling bills. In Salt Lake City, Utah, where temperatures on dark rooftops can soar to 65°C, the rooftop of a large warehouse was painted white. "I've been up on it plenty of times," says a worker at the warehouse. "It doesn't come up and just drill you with heat like the black ones do."

Planting even small bushes and trees can help cool a city.

CONNECTIONS Design and Research Go to the Glencoe Science Web site to research NASA's Urban Heat Island Project. What actions are cities taking to reduce the heat-island effect? Design a city area that would help reduce this effect.

Reviewing Main Ideas

Section 1 Temperature and Thermal Energy

1. Molecules of matter are moving constantly. Temperature is related to the average value of the kinetic energy of the molecules.

2. Thermometers measure temperature. Three common temperature scales are the Celsius, Fahrenheit, and Kelvin scales.

3. Thermal energy is the total kinetic and potential energy of the particles in matter. *How has thermal energy changed when this iron has melted?*

Section 2 Heat

1. Heat is thermal energy that is transferred from a warmer object to a colder object.

2. Heat can be transferred by conduction, convection, and radiation. *Why do you feel warm when you stand in front of a fireplace?*

3. A material that easily transfers heat is called a conductor. A material that resists the flow of heat is an insulator.

4. The specific heat of a substance is the amount of heat needed to change the temperature of 1 kg of the substance 1°C.

5. Thermal pollution occurs when warm water is added to a body of water, such as a river or lake.

Section 3 Engines and Refrigerators

1. A device that converts thermal energy into mechanical energy is an engine.

2. In an internal combustion engine, fuel is burned in combustion chambers inside the engine.

3. Internal combustion engines that are used in cars and airplanes burn fuel to do work, using a four-stroke cycle.

4. Refrigerators and air conditioners use a coolant to move heat from one place to another. *Why is one side of this air conditioner placed outdoors?*

FOLDABLES
Reading & Study Skills

After You Read

Write what you learned about the relationship between heat and thermal energy on the back of your Foldable.

Visualizing Main Ideas

Complete the following cycle map about the four-stroke cycle.

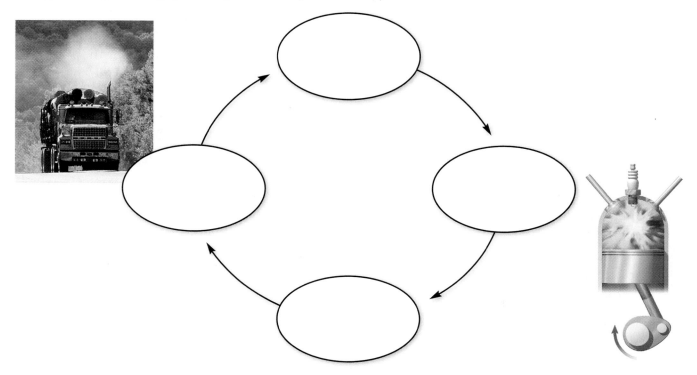

Vocabulary Review

Vocabulary Words

a. conduction
b. conductor
c. convection
d. engine
e. heat
f. internal combustion engine

g. radiation
h. specific heat
i. temperature
j. thermal energy
k. thermal pollution

Using Vocabulary

Explain the differences in the vocabulary words given below. Then explain how the words are related. Use complete sentences in your answers.

1. internal combustion engine, engine
2. temperature, thermal energy
3. thermal energy, thermal pollution
4. conduction, convection
5. conduction, heat
6. heat, specific heat
7. conduction, radiation
8. convection, radiation
9. conductor, heat

Study Tip

Practice reading tables. See whether you can devise a graph that shows the same information that a table shows.

Chapter 6 Assessment

Checking Concepts

Choose the word or phrase that best answers the question.

1. What source of thermal energy does an internal combustion engine use?
   A) steam
   B) hot water
   C) burning fuel
   D) refrigerant

2. What happens to most materials when they become warmer?
   A) They contract.
   B) They float.
   C) They vaporize.
   D) They expand.

3. Which type of heat transfer occurs when two objects at different temperatures are touching?
   A) convection
   B) radiation
   C) condensation
   D) conduction

4. Which of the following describes the thermal energy of particles in a substance?
   A) average value of all kinetic energy
   B) total value of all kinetic energy
   C) total value of all kinetic and potential energy
   D) average value of all kinetic and potential energy

5. Heat being transferred from the Sun to Earth is an example of which process?
   A) convection
   B) expansion
   C) radiation
   D) conduction

6. Many insulating materials contain spaces filled with air because air is what type of material?
   A) conductor
   B) coolant
   C) radiator
   D) insulator

7. What do thermometers measure?
   A) average kinetic energy of particles
   B) heat of particles
   C) evaporation rate
   D) total energy of particles

8. Which of the following is true?
   A) Warm air is less dense than cool air.
   B) Warm air is as dense as cool air.
   C) Warm air has no density.
   D) Warm air is denser than cool air.

9. Which of these is the name for thermal energy that moves from a warmer object to a cooler one?
   A) kinetic energy
   B) specific heat
   C) heat
   D) temperature

10. If the same amounts of heat were added to equal masses of the following objects, which would get hottest?
    A) object with low specific heat
    B) object with medium specific heat
    C) object with high specific heat
    D) object with very high specific heat

Thinking Critically

11. Water is a poor conductor of heat. Yet, when you heat water in a pan, the surface gets hot quickly, even though you are applying heat to the bottom of the water. Explain.

12. List the following temperatures from coldest to warmest: 80° C, 200 K, 50° F.

13. Why do several layers of clothing often keep you warmer than a single layer?

14. The phrase "heat rises" is sometimes used to describe the behavior of heat. For what type of materials is this phrase correct? Explain.

15. In a refrigerator, the coolant absorbs heat from inside the refrigerator and then transfers this heat to the air outside. Describe how the temperature of the coolant is changed as it flows through the refrigerator.

Developing Skills

16. Designing an Experiment Some colors of clothing absorb heat better than other colors. Design an experiment that will test various colors by placing them in the hot Sun for a period of time. Explain your results.

17. Drawing Conclusions Would it be possible to cool a kitchen by leaving the refrigerator door open? Explain.

18. Concept Mapping Complete the following concept map on convection in a liquid.

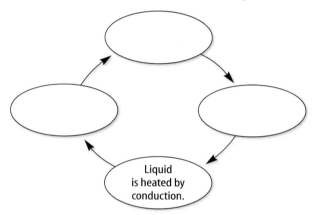

Liquid is heated by conduction.

Performance Assessment

19. Poll In the United States, the Fahrenheit temperature scale is used most often. Some people feel that Americans should switch to the Celsius scale. Take a poll of at least 20 people. Find out if they feel the switch to Celsius should be made. Make a list of reasons people give for or against changing.

TECHNOLOGY

Go to the Glencoe Science Web site at **science.glencoe.com** or use the **Glencoe Science CD-ROM** for additional chapter assessment.

THE PRINCETON REVIEW — Test Practice

Mrs. Keeley's chemistry class is studying how ice changes to water and then to steam as it is heated. They measured the amount of heat added, and the temperature of the ice, water, and steam. The graph below shows their results.

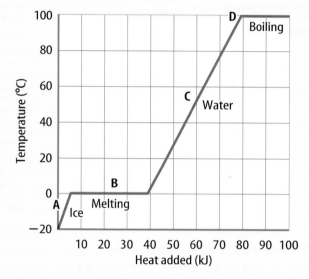

Study the graph and answer the following questions.

1. According to the graph above, at what temperature does ice change to water?
A) 10° C **C)** 0° C
B) 100° C **D)** −10° C

2. Temperature is a measure of the average kinetic energy of the molecules of a substance. Over what part of the graph does the kinetic energy of the molecules stay the same?
F) A
G) C
H) B
J) Not enough information given

Reading Comprehension

Read the passage. Then read each question that follows the passage. Decide which is the best answer to each question.

Test-Taking Tip Make a list of the important details in the passage.

Bouncing Back

Have you ever noticed that the balls you use for different sports bounce differently? If you played baseball with a tennis ball, the ball probably would fly into the outfield without much effort when you hit it with your bat. In contrast, if you used a baseball in a tennis match, the ball probably would not bounce high enough for your opponent to hit it very well. The difference in the way balls bounce depends upon the materials that make up the balls and the way they are constructed.

A ball drops to the floor as a result of gravity. As the ball drops, it gathers speed. When the ball hits the floor, the energy that it has gained goes into deforming the ball, changing it from its round shape. As the ball changes shape, the molecules within it stretch farther apart in some places and squeeze closer together in other places. The strength of the bonds between molecules determines how much they stretch apart and squeeze together. This depends on the chemical composition of the materials in the ball.

Most balls are made of rubber. Rubber is *elastic,* which means that it returns to its original shape after it's been deformed. Rubber is made of molecules called polymers that are long chains. Normally these chains are coiled up, but when the rubber is stretched, the chains straighten out. Then when the stretching force is removed, the chains coil up again. How high a ball bounces depends on the type of polymer molecules the rubber is made of. Bouncing balls sometimes feel warm because some of the ball's kinetic energy is converted into thermal energy.

A baseball is struck by a bat and flattens. The ball bounces off the bat as it becomes round again.

1. According to information in the passage, it is probably accurate to conclude that _____.
   A) all rubber balls bounce the same, no matter what they are made of
   B) the way rubber balls bounce depends upon the polymers that they are made of
   C) baseballs are better for playing tennis than tennis balls are for playing baseball
   D) the higher a ball bounces, the more thermal energy is produced

2. In the context of this passage, the word <u>elastic</u> means _____.
   F) able to retain its shape
   G) inflexible
   H) tightly linked
   J) warm

Reasoning and Skills

Read each question and choose the best answer.

1. Why aren't you doing any work when you push against a solid brick wall?
   A) because the wall is much heavier than you are
   B) because you can exert only a very small force
   C) because only machines can do work
   D) because you haven't moved the wall

Test-Taking Tip Recall the definition of work.

2. Which of the following is true about this simple machine?
   F) It increases the force needed to lift the object.
   G) It decreases the force needed to lift the object.
   H) It decreases the work needed to lift the object.
   J) It decreases the object's weight.

Test-Taking Tip Consider how the machine works and what you need it to do.

3. Which of the following statements is the best explanation for what likely happened to the sidewalk in the picture above?
   A) An animal that lives underground has tunneled under the sidewalk.
   B) The sidewalk has expanded because of extremely hot weather.
   C) The sidewalk has heaved due to the number of people and other animals that have walked on it.
   D) The sidewalk has contracted because of extremely hot weather.

Test-Taking Tip Consider the kind of force needed to cause such a break. How could this force most likely be produced?

Consider the question carefully before writing your answer on a separate sheet of paper.

4. You have 2 L of water at a temperature of 5°C and 1/4 L of water at a temperature of 5°C. Which volume has more thermal energy? Why?

Test-Taking Tip Think carefully about the definition of thermal energy.

Student Resources

Student Resources

CONTENTS

Field GUIDE

If you like smooth, gentle rides, don't expect to get one at an amusement park. Amusement park rides are designed to provide thrills—plummeting down hills at 160 km/h, whizzing around curves so fast you think you'll fall out of your seat, zooming upside down, plunging over waterfalls, dropping so fast and far that you feel weightless. It's all part of the fun.

May the Force Be with You

What you might not realize as you're screaming with delight is that amusement park rides are lessons in physics. You can apply Newton's laws of motion to everything from the water slide and the bumper cars to the roller coasters. Amusement park ride designers know how to use the laws of motion to jolt, bump, and jostle you enough to make you scream, while still keeping you safe from harm. They don't just plan how the laws of motion cause these rides to move, they also plan how you will move when you are on the rides. These designers also use Newton's law of motion when they design and build the rides to make the structures safe and lasting. Look at the forces at work on some popular amusement park rides.

Amusement Park Rides

Free-fall ride

Free Fall

Slowly you rise up, up, up. Gravity is pulling you downward, but your seat exerts an upward force on you. Then, in an instant, you're plummeting toward the ground at speeds of more than 100 km/h. When you fall, your seat falls at the same rate and no longer exerts a force on you. Because you don't feel your seat pushing upward, you have the feeling of being weightless—at least for a few seconds.

Field Activity

The next time you're at an amusement park, watch the rides. When you return home, make drawings of the rides using arrows to show how they move. Group the rides according to their movements. Compare your drawings and observations to the information in this field guide.

Roller Coaster: Design

The biggest coasters—some as tall as a 40-story building—are made of steel. Steel roller coasters are stronger and sway less than wooden roller coasters. This allows for more looping, more hills, and faster speeds.

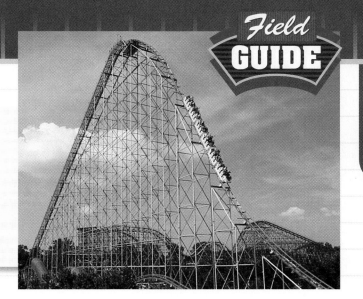

Roller coaster

Roller Coaster: The Coaster's Motion

Roller coasters are gravity-powered trains. Some coasters have motor-driven chains that move the cars to the top of the first hill. Then, gravity keeps it going.

The first hill is the highest point on the track. As the coaster rolls down the first hill, it converts potential energy to kinetic energy that sends it up the next hill. With each hill it climbs, it loses a little energy due to friction. That is why each hill is generally lower than the one before it.

Roller Coaster: Your Ride

Inertia is at work when you sweep around curves on a roller coaster. Inertia is the tendency for a body that's moving in a certain direction to keep moving in the same direction. For example, when the coaster swings right, inertia tries to keep you going in a straight line at a constant speed. As a result, you are pushed to the left side of your car.

Inertia tends to keep bodies moving in a straight line.

Bumper Cars: The Car's Motion

You control your bumper car's acceleration with the accelerator pedal. When the car you're in bumps head-on into another car, your car comes to an abrupt stop. The big rubber bumper around the bottom of the car diffuses the force of the collision by prolonging the impact.

Bumper Cars: Your Ride

When you first accelerate in a bumper car, you feel as though you are being pushed back in your seat. This sensation and the jolt you feel when you hit another car are due to inertia. On impact, your car stops, but your inertia makes you continue to move forward. It's the same jolt you feel in a car when someone slams on the brakes.

In a bumper-car collision, inertia keeps each rider moving forward.

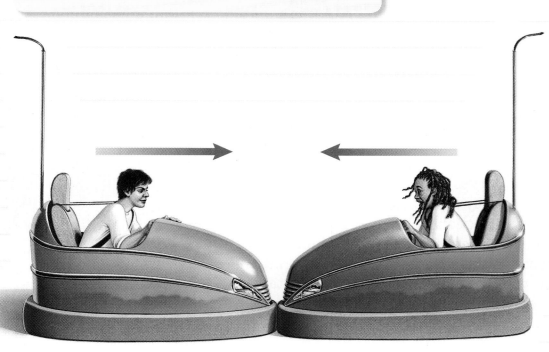

Swing Ride: Design

Some of the more powerful swing rides make about eight revolutions around the central pole each minute. These swing rides are capable of moving their riders at speeds of close to 50 km/h.

The arrows show the forces at work in a swing ride.

Swing Ride: Forces

As the swings rotate, your inertia wants to fling you outward, but the chain that connects your seat to the ride's central pole prevents you from being flung into the air. You can see the changes in force as the swing ride changes speeds. As the ride speeds up and the forces exerted on the chain increase, your swing rises, moves outward, and travels almost parallel to the ground. As the ride slows, these forces on the chains decrease, returning the swings slowly to their original position.

Field GUIDE

Early astronauts were crammed into tiny space capsules where they could barely move in their seats. Food was a tasteless paste squeezed from a tube or a hard, bite-sized cube. Today, space shuttle astronauts have a two-level cabin with sleeping bunks, a galley for preparing food, and exercise equipment. Living in space isn't what it used to be.

Living in Orbit

Although conditions on a spacecraft are better now than in the past, the problems astronauts face are the same. They still go about their daily life, but space has no air, food, or water. This makes it hard to prepare meals and wash dishes afterward. It complicates how you drink beverages out of a glass. Due to these challenges, space shuttle crews must carry everything they need with them to survive in space.

By far the biggest challenge for astronauts is still the lack of gravity. Imagine eating a meal as part of it floats away, or sleeping in a bed that drifts into walls. NASA scientists have found ways to overcome such problems. This field guide offers a look at some of them.

Living in Space

Life-Support System

People need oxygen to breathe. The shuttle carries canisters of super-cold liquid oxygen and pressurized nitrogen to create an atmosphere in the crew compartment that is similar to Earth's— 79 percent nitrogen and 21 percent oxygen. The shuttle also circulates air through canisters of lithium hydroxide and activated charcoal, removing carbon dioxide and odors from it. Crew members must change one of the two canisters every 12 h.

Field Activity

Read a science-fiction description of people living and working in space. In your Science Journal, describe how people performed daily tasks such as eating, sleeping, and getting around. Go to the Glencoe Science Web site at **science.glencoe.com** and click on the NASA link to find out more about living and working in space. Compare what you wrote with what you learn in this field guide.

Electricity

Fuel cells generate electricity by chemically combining hydrogen and oxygen. As a by-product, they produce 3 kg of water each hour—some of which is used to prepare food.

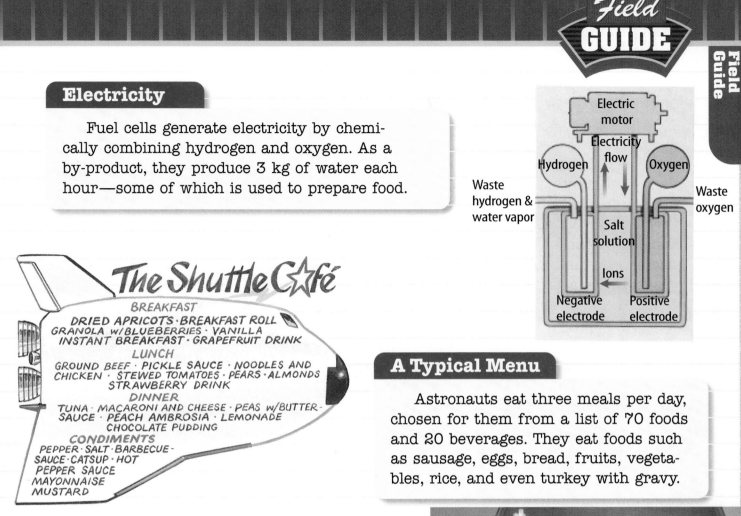

The Shuttle Café

BREAKFAST
DRIED APRICOTS · BREAKFAST ROLL
GRANOLA w/BLUEBERRIES · VANILLA
INSTANT BREAKFAST · GRAPEFRUIT DRINK

LUNCH
GROUND BEEF · PICKLE SAUCE · NOODLES AND
CHICKEN · STEWED TOMATOES · PEARS · ALMONDS
STRAWBERRY DRINK

DINNER
TUNA · MACARONI AND CHEESE · PEAS w/BUTTER-
SAUCE · PEACH AMBROSIA · LEMONADE
CHOCOLATE PUDDING

CONDIMENTS
PEPPER · SALT · BARBECUE-
SAUCE · CATSUP · HOT
PEPPER SAUCE
MAYONNAISE
MUSTARD

A Typical Menu

Astronauts eat three meals per day, chosen for them from a list of 70 foods and 20 beverages. They eat foods such as sausage, eggs, bread, fruits, vegetables, rice, and even turkey with gravy.

Food Preservation

Foods are not refrigerated. Some foods are freeze-dried, so water is added before they are eaten. Some foods are heated to kill bacteria and sealed in airtight foil packets. Irradiated food, such as bread and some meat, has been exposed to radiation to kill bacteria.

Food Preparation

Astronauts prepare and eat their food in the galley. A different person serves each meal, which takes about 20 min to prepare. The astronaut injects water into dried or powdered foods that need it, and puts hot dishes into the oven to warm them. Some foods can be eaten right out of the pouches.

These astronauts are enjoying a meal together.

Working Out

To help prevent bone and muscle deterioration due to space's weightless environment, astronauts exercise for 15 min each day on 7-day to 14-day missions. They work out for 30 min daily on 30-day missions. They can use a treadmill, a rowing machine, or an exercise bike. Even with this exercise, astronauts can lose more than one percent of their bone density for each month they are in space.

Using a rowing machine

Exercise Equipment

The base of the treadmill hooks into the floor or walls. An astronaut can stand on the treadmill with rubber bungee cords attached to a belt and shoulder harness. The cord is tightened to increase resistance.

Using a treadmill

Getting Some Sleep

Weightless astronauts can sleep in unusual places. Each astronaut's sleep station contains a bed made up of a padded board with a fireproof sleeping bag attached. Two astronauts sleep on bunks facing up. One sleeps on the underside of the lower bunk, facing the floor. The fourth sleeps vertically against the wall.

Sleeping compartments

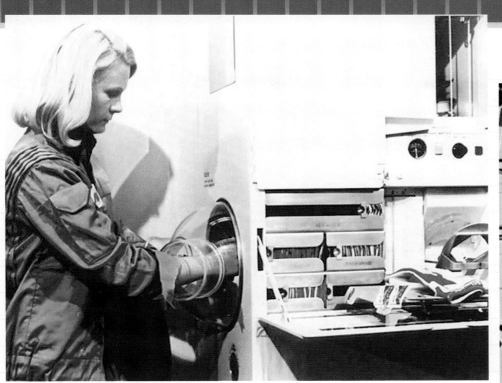

This is their hand-washing station.

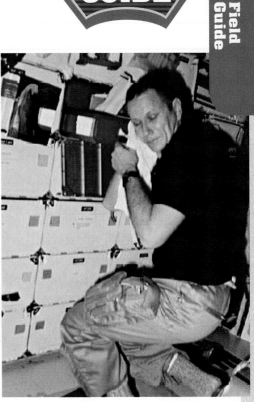

This astronaut uses a wet cloth to keep clean.

Cleaning Up

After 8 h of sleep, astronauts have 45 min for morning hygiene. There aren't any showers or baths in space. To keep clean, astronauts just wipe themselves (and their hair) off with a wet cloth. They also can wash their hands at the hand-washing station. Water is air-blasted at their hands and then immediately sucked up.

Waste Management

Astronauts have a special toilet they use in space. It utilizes air instead of water to remove bodily wastes. The waste is then held in a tank until the spacecraft returns to Earth.

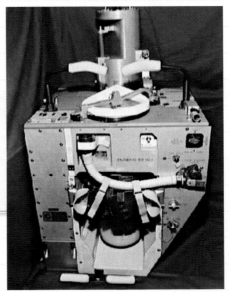

Here is a space shuttle toilet.

Organizing Information

As you study science, you will make many observations and conduct investigations and experiments. You will also research information that is available from many sources. These activities will involve organizing and recording data. The quality of the data you collect and the way you organize it will determine how well others can understand and use it. In **Figure 1,** the student is obtaining and recording information using a thermometer.

Putting your observations in writing is an important way of communicating to others the information you have found and the results of your investigations and experiments.

Researching Information

Scientists work to build on and add to human knowledge of the world. Before moving in a new direction, it is important to gather the information that already is known about a subject. You will look for such information in various reference sources. Follow these steps to research information on a scientific subject:

Step 1 Determine exactly what you need to know about the subject. For instance, you might want to find out about one of the elements in the periodic table.

Step 2 Make a list of questions, such as: Who discovered the element? When was it discovered? What makes the element useful or interesting?

Step 3 Use multiple sources such as textbooks, encyclopedias, government documents, professional journals, science magazines, and the Internet.

Step 4 List where you found the sources. Make sure the sources you use are reliable and the most current available.

Figure 1
Making an observation is one way to gather information directly.

Evaluating Print and Nonprint Sources

Not all sources of information are reliable. Evaluate the sources you use for information, and use only those you know to be dependable. For example, suppose you want to find ways to make your home more energy efficient. You might find two Web sites on how to save energy in your home. One Web site contains "Energy-Saving Tips" written by a company that sells a new type of weatherproofing material you put around your door frames. The other is a Web page on "Conserving Energy in Your Home" written by the U.S. Department of Energy. You would choose the second Web site as the more reliable source of information.

In science, information can change rapidly. Always consult the most current sources. A 1985 source about saving energy would not reflect the most recent research and findings.

Interpreting Scientific Illustrations

As you research a science topic, you will see drawings, diagrams, and photographs. Illustrations help you understand what you read. Some illustrations are included to help you understand an idea that you can't see easily by yourself. For instance, you can't see the tiny particles in an atom, but you can look at a diagram of an atom as labeled in **Figure 2** that helps you understand something about it. Visualizing a drawing helps many people remember details more easily. Illustrations also provide examples that clarify difficult concepts or give additional information about the topic you are studying.

Most illustrations have a label or caption. A label or caption identifies the illustration or provides additional information to better explain it. Can you find the caption or labels in **Figure 2?**

Figure 2
This drawing shows an atom of carbon with its six protons, six neutrons, and six electrons.

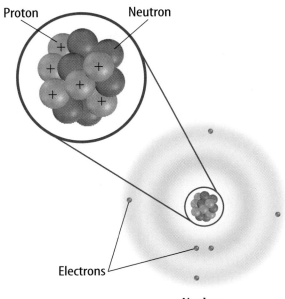

Venn Diagram

Although it is not a concept map, a Venn diagram illustrates how two subjects compare and contrast. In other words, you can see the characteristics that the subjects have in common and those that they do not.

The Venn diagram in **Figure 3** shows the relationship between two different substances made from the element carbon. However, due to the way their atoms are arranged, one substance is the gemstone diamond, and the other is the graphite found in pencils.

Concept Mapping

If you were taking a car trip, you might take some sort of road map. By using a map, you begin to learn where you are in relation to other places on the map.

A concept map is similar to a road map, but a concept map shows relationships among ideas (or concepts) rather than places. It is a diagram that visually shows how concepts are related. Because a concept map shows relationships among ideas, it can make the meanings of ideas and terms clear and help you understand what you are studying.

Overall, concept maps are useful for breaking large concepts down into smaller parts, making learning easier.

Figure 3
A Venn diagram shows how objects or concepts are alike and how they are different.

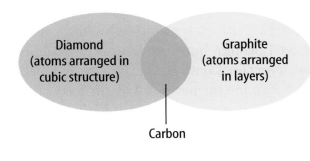

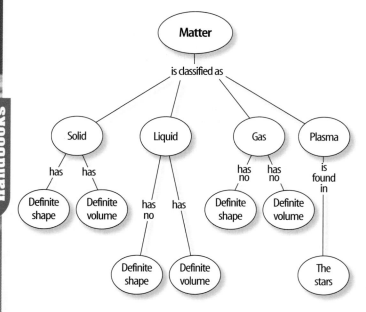

Figure 4
A network tree shows how concepts or objects are related.

Network Tree Look at the network tree in **Figure 4,** that describes the different types of matter. A network tree is a type of concept map. Notice how some words are in ovals while others are written across connecting lines. The words inside the ovals are science terms or concepts. The words written on the connecting lines describe the relationships between the concepts.

When constructing a network tree, write the topic on a note card or piece of paper. Write the major concepts related to that topic on separate note cards or pieces of paper. Then arrange them in order from general to specific. Branch the related concepts from the major concept and describe the relationships on the connecting lines. Continue branching to more specific concepts. If necessary, write the relationships between the concepts on the connecting lines until all concepts are mapped. Then examine the network tree for relationships that cross branches, and add them to the network tree.

Events Chain An events chain is another type of concept map. It models the order, or sequence, of items. In science, an events chain can be used to describe a sequence of events, the steps in a procedure, or the stages of a process.

When making an events chain, first find the one event that starts the chain. This event is called the initiating event. Then, find the next event in the chain and continue until you reach an outcome. Suppose you are asked to describe why and how a sound might make an echo. You might draw an events chain such as the one in **Figure 5.** Notice that connecting words are not necessary in an events chain.

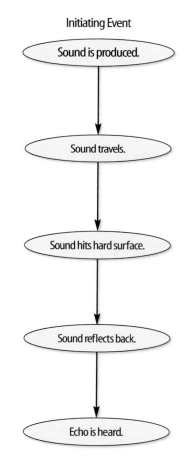

Figure 5
Events chains show the order of steps in a process or event.

Cycle Map A cycle concept map is a specific type of events chain map. In a cycle concept map, the series of events does not produce a final outcome. Instead, the last event in the chain relates back to the beginning event.

You first decide what event will be used as the beginning event. Once that is decided, you list events in order that occur after it. Words are written between events that describe what happens from one event to the next. The last event in a cycle concept map relates back to the beginning event. The number of events in a cycle concept varies, but is usually three or more. Look at the cycle map, as shown in **Figure 6.**

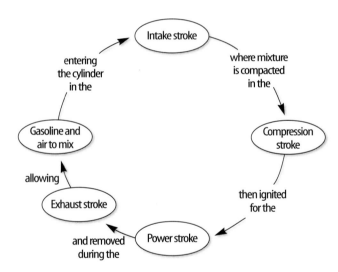

Figure 6
A cycle map shows events that occur in a cycle.

Spider Map A type of concept map that you can use for brainstorming is the spider map. When you have a central idea, you might find you have a jumble of ideas that relate to it but are not necessarily clearly related to each other. The spider map on sound in **Figure 7** shows that if you write these ideas outside the main concept, then you can begin to separate and group un-related terms so they become more useful.

Figure 7
A spider map allows you to list ideas that relate to a central topic but not necessarily to one another.

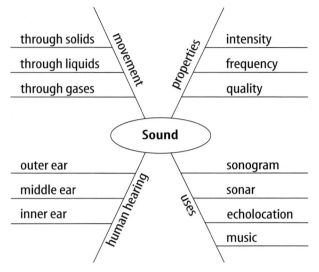

Writing a Paper

You will write papers often when researching science topics or reporting the results of investigations or experiments. Scientists frequently write papers to share their data and conclusions with other scientists and the public. When writing a paper, use these steps.

Step 1 Assemble your data by using graphs, tables, or a concept map. Create an outline.

Step 2 Start with an introduction that contains a clear statement of purpose and what you intend to discuss or prove.

Step 3 Organize the body into paragraphs. Each paragraph should start with a topic sentence, and the remaining sentences in that paragraph should support your point.

Step 4 Position data to help support your points.

Step 5 Summarize the main points and finish with a conclusion statement.

Step 6 Use tables, graphs, charts, and illustrations whenever possible.

You might say the work of a scientist is to solve problems. When you decide to find out why your neighbor's hydrangeas produce blue flowers while yours are pink, you are problem solving, too. You might also observe that your neighbor's azaleas are healthier than yours are and decide to see whether differences in the soil explain the differences in these plants.

Scientists use orderly approaches to solve problems. The methods scientists use include identifying a question, making observations, forming a hypothesis, testing a hypothesis, analyzing results, and drawing conclusions.

Scientific investigations involve careful observation under controlled conditions. Such observation of an object or a process can suggest new and interesting questions about it. These questions sometimes lead to the formation of a hypothesis. Scientific investigations are designed to test a hypothesis.

Identifying a Question

The first step in a scientific investigation or experiment is to identify a question to be answered or a problem to be solved. You might be interested in knowing how beams of laser light like the ones in **Figure 8** look the way they do.

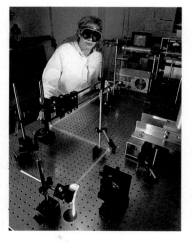

Figure 8
When you see lasers being used for scientific research, you might ask yourself, "Are these lasers different from those that are used for surgery?"

Forming Hypotheses

Hypotheses are based on observations that have been made. A hypothesis is a possible explanation based on previous knowledge and observations.

Perhaps a scientist has observed that certain substances dissolve faster in warm water than in cold. Based on these observations, the scientist can make a statement that he or she can test. The statement is a hypothesis. The hypothesis could be: *A substance dissolves in warm water faster.* A hypothesis has to be something you can test by using an investigation. A testable hypothesis is a valid hypothesis.

Predicting

When you apply a hypothesis to a specific situation, you predict something about that situation. First, you must identify which hypothesis fits the situation you are considering. People use predictions to make everyday decisions. Based on previous observations and experiences, you might form a prediction that if substances dissolve in warm water faster, then heating the water will shorten mixing time for powdered fruit drinks. Someone could use this prediction to save time in preparing a fruit punch for a party.

Testing a Hypothesis

To test a hypothesis, you need a procedure. A procedure is the plan you follow in your experiment. A procedure tells you what materials to use, as well as how and in what order to use them. When you follow a procedure, data are generated that support or do not support the original hypothesis statement.

For example, premium gasoline costs more than regular gasoline. Does premium gasoline increase the efficiency or fuel mileage of your family car? You decide to test the hypothesis: "If premium gasoline is more efficient, then it should increase the fuel mileage of my family's car." Then you write the procedure shown in **Figure 9** for your experiment and generate the data presented in the table below.

Figure 9
A procedure tells you what to do step by step.

> ## Procedure
> 1. Use regular gasoline for two weeks.
> 2. Record the number of kilometers between fill-ups and the amount of gasoline used.
> 3. Switch to premium gasoline for two weeks.
> 4. Record the number of kilometers between fill-ups and the amount of gasoline used.

Gasoline Data			
Type of Gasoline	Kilometers Traveled	Liters Used	Liters per Kilometer
Regular	762	45.34	0.059
Premium	661	42.30	0.064

These data show that premium gasoline is less efficient than regular gasoline in one particular car. It took more gasoline to travel 1 km (0.064) using premium gasoline than it did to travel 1 km using regular gasoline (0.059). This conclusion does not support the hypothesis.

Are all investigations alike? Keep in mind as you perform investigations in science that a hypothesis can be tested in many ways. Not every investigation makes use of all the ways that are described on these pages, and not all hypotheses are tested by investigations. Scientists encounter many variations in the methods that are used when they perform experiments. The skills in this handbook are here for you to use and practice.

Identifying and Manipulating Variables and Controls

In any experiment, it is important to keep everything the same except for the item you are testing. The one factor you change is called the independent variable. The factor that changes as a result of the independent variable is called the dependent variable. Always make sure you have only one independent variable. If you allow more than one, you will not know what causes the changes you observe in the dependent variable. Many experiments also have controls—individual instances or experimental subjects for which the independent variable is not changed. You can then compare the test results to the control results.

For example, in the fuel-mileage experiment, you made everything the same except the type of gasoline that was used. The driver, the type of automobile, and the type of driving were the same throughout. In this way, you could be sure that any mileage differences were caused by the type of fuel—the independent variable. The fuel mileage was the dependent variable.

If you could repeat the experiment using several automobiles of the same type on a standard driving track with the same driver, you could make one automobile a control by using regular gasoline over the four-week period.

Collecting Data

Whether you are carrying out an investigation or a short observational experiment, you will collect data, or information. Scientists collect data accurately as numbers and descriptions and organize it in specific ways.

Observing Scientists observe items and events, then record what they see. When they use only words to describe an observation, it is called qualitative data. For example, a scientist might describe the color, texture, or odor of a substance produced in a chemical reaction. Scientists' observations also can describe how much there is of something. These observations use numbers, as well as words, in the description and are called quantitative data. For example, if a sample of the element gold is described as being "shiny and very dense," the data are clearly qualitative. Quantitative data on this sample of gold might include "a mass of 30 g and a density of 19.3 g/cm³." Quantitative data often are organized into tables. Then, from information in the table, a graph can be drawn. Graphs can reveal relationships that exist in experimental data.

When you make observations in science, you should examine the entire object or situation first, then look carefully for details. If you're looking at an element sample, for instance, check the general color and pattern of the sample before using a hand lens to examine its surface for any smaller details or characteristics. Remember to record accurately everything you see.

Scientists try to make careful and accurate observations. When possible, they use instruments such as microscopes, metric rulers, graduated cylinders, thermometers, and balances. Measurements provide numerical data that can be repeated and checked.

Sampling When working with large numbers of objects or a large population, scientists usually cannot observe or study every one of them. Instead, they use a sample or a portion of the total number. To *sample* is to take a small, representative portion of the objects or organisms of a population for research. By making careful observations or manipulating variables within a portion of a group, information is discovered and conclusions are drawn that might apply to the whole population.

Estimating Scientific work also involves estimating. To estimate is to make a judgment about the size or the number of something without measuring or counting every object or member of a population. Scientists first measure or count the amount or number in a small sample. A geologist, for example, might remove a 10-g sample from a large rock that is rich in copper ore, as in **Figure 10.** Then a chemist would determine the percentage of copper by mass and multiply that percentage by the total mass of the rock to estimate the total mass of copper in the large rock.

Figure 10
Determining the percentage of copper by mass that is present in a small piece of a large rock, which is rich in copper ore, can help estimate the total mass of copper ore that is present in the rock.

Measuring in SI

The metric system of measurement was developed in 1795. A modern form of the metric system, called the International System, or SI, was adopted in 1960. SI provides standard measurements that all scientists around the world can understand.

The metric system is convenient because unit sizes vary by multiples of 10. When changing from smaller units to larger units, divide by a multiple of 10. When changing from larger units to smaller, multiply by a multiple of 10. To convert millimeters to centimeters, divide the millimeters by 10. To convert 30 mm to centimeters, divide 30 by 10 (30 mm equal 3 cm).

Prefixes are used to name units. Look at the table below for some common metric prefixes and their meanings. Do you see how the prefix *kilo-* attached to the unit *gram* is *kilogram*, or 1,000 g?

Metric Prefixes			
Prefix	Symbol	Meaning	
kilo-	k	1,000	thousand
hecto-	h	100	hundred
deka-	da	10	ten
deci-	d	0.1	tenth
centi-	c	0.01	hundredth
milli-	m	0.001	thousandth

Now look at the metric ruler shown in **Figure 11.** The centimeter lines are the long, numbered lines, and the shorter lines are millimeter lines.

When using a metric ruler, line up the 0-cm mark with the end of the object being measured, and read the number of the unit where the object ends, in this instance it would be 4.5 cm.

Figure 11
This metric ruler has centimeter and millimeter divisions.

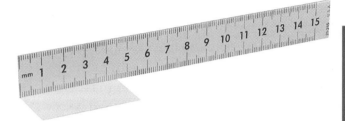

Liquid Volume In some science activities, you will measure liquids. The unit that is used to measure liquids is the liter. A liter has the volume of 1,000 cm³. The prefix *milli-* means "thousandth (0.001)." A milliliter is one thousandth of 1 L, and 1 L has the volume of 1,000 mL. One milliliter of liquid completely fills a cube measuring 1 cm on each side. Therefore, 1 mL equals 1 cm³.

You will use beakers and graduated cylinders to measure liquid volume. A graduated cylinder, as illustrated in **Figure 12,** is marked from bottom to top in milliliters. This one contains 79 mL of a liquid.

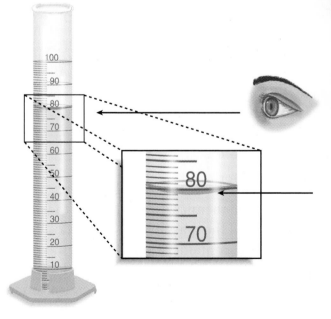

Figure 12
Graduated cylinders measure liquid volume.

Skill Handbooks

Mass Scientists measure mass in grams. You might use a beam balance similar to the one shown in **Figure 13.** The balance has a pan on one side and a set of beams on the other side. Each beam has a rider that slides on the beam.

Before you find the mass of an object, slide all the riders back to the zero point. Check the pointer on the right to make sure it swings an equal distance above and below the zero point. If the swing is unequal, find and turn the adjusting screw until you have an equal swing.

Place an object on the pan. Slide the largest rider along its beam until the pointer drops below zero. Then move it back one notch. Repeat the process on each beam until the pointer swings an equal distance above and below the zero point. Sum the masses on each beam to find the mass of the object. Move all riders back to zero when finished.

Figure 13
A triple beam balance is used to determine the mass of an object.

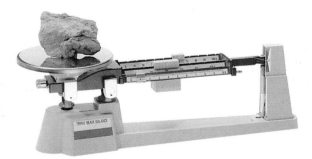

You should never place a hot object on the pan or pour chemicals directly onto the pan. Instead, find the mass of a clean container. Remove the container from the pan, then place the chemicals in the container. Find the mass of the container with the chemicals in it. To find the mass of the chemicals, subtract the mass of the empty container from the mass of the filled container.

Making and Using Tables

Browse through your textbook and you will see tables in the text and in the activities. In a table, data, or information, are arranged so that they are easier to understand. Activity tables help organize the data you collect during an activity so results can be interpreted.

Making Tables To make a table, list the items to be compared in the first column and the characteristics to be compared in the first row. The title should clearly indicate the content of the table, and the column or row heads should tell the reader what information is found in there. The table below lists materials collected for recycling on three weekly pick-up days. The inclusion of kilograms in parentheses also identifies for the reader that the figures are mass units.

Recyclable Materials Collected During Week			
Day of Week	Paper (kg)	Aluminum (kg)	Glass (kg)
Monday	5.0	4.0	12.0
Wednesday	4.0	1.0	10.0
Friday	2.5	2.0	10.0

Using Tables How much paper, in kilograms, is being recycled on Wednesday? Locate the column labeled "Paper (kg)" and the row "Wednesday." The information in the box where the column and row intersect is the answer. Did you answer "4.0"? How much aluminum, in kilograms, is being recycled on Friday? If you answered "2.0," you understand how to read the table. How much glass is collected for recycling each week? Locate the column labeled "Glass (kg)" and add the figures for all three rows. If you answered "32.0," then you know how to locate and use the data provided in the table.

Recording Data

To be useful, the data you collect must be recorded carefully. Accuracy is key. A well-thought-out experiment includes a way to record procedures, observations, and results accurately. Data tables are one way to organize and record results. Set up the tables you will need ahead of time so you can record the data right away.

Record information properly and neatly. Never put unidentified data on scraps of paper. Instead, data should be written in a notebook like the one in **Figure 14.** Write in pencil so information isn't lost if your data get wet. At each point in the experiment, record your information and label it. That way, your data will be accurate and you will not have to determine what the figures mean when you look at your notes later.

Figure 14
Record data neatly and clearly so they are easy to understand.

Recording Observations

It is important to record observations accurately and completely. That is why you always should record observations in your notes immediately as you make them. It is easy to miss details or make mistakes when recording results from memory. Do not include your personal thoughts when you record your data. Record only what you observe to eliminate bias. For example, when you record the time required for five students to climb the same set of stairs, you would note which student took the longest time. However, you would not refer to that student's time as "the worst time of all the students in the group."

Making Models

You can organize the observations and other data you collect and record in many ways. Making models is one way to help you better understand the parts of a structure you have been observing or the way a process for which you have been taking various measurements works.

Models often show things that are too large or too small for normal viewing. For example, you normally won't see the inside of an atom. However, you can understand the structure of the atom better by making a three-dimensional model of an atom. The relative sizes, the positions, and the movements of protons, neutrons, and electrons can be explained in words. An atomic model made of a plastic-ball nucleus and pipe-cleaner electron shells can help you visualize how the parts of the atom relate to each other.

Other models can be devised on a computer. Some models, such as those that illustrate the chemical combinations of different elements, are mathematical and are represented by equations.

Making and Using Graphs

After scientists organize data in tables, they might display the data in a graph that shows the relationship of one variable to another. A graph makes interpretation and analysis of data easier. Three types of graphs are the line graph, the bar graph, and the circle graph.

Line Graphs A line graph like in **Figure 15** is used to show the relationship between two variables. The variables being compared go on two axes of the graph. For data from an experiment, the independent variable always goes on the horizontal axis, called the *x*-axis. The dependent variable always goes on the vertical axis, called the *y*-axis. After drawing your axes, label each with a scale. Next, plot the data points.

A data point is the intersection of the recorded value of the dependent variable for each tested value of the independent variable. After all the points are plotted, connect them.

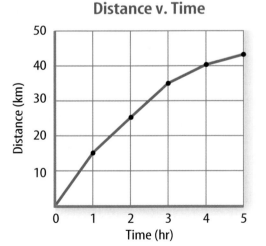

Figure 15
This line graph shows the relationship between distance and time during a bicycle ride lasting several hours.

Bar Graphs Bar graphs compare data that do not change continuously. Vertical bars show the relationships among data.

To make a bar graph, set up the *y*-axis as you did for the line graph. Draw vertical bars of equal size from the *x*-axis up to the point on the *y*-axis that represents the value of *x*.

Figure 16
The amount of aluminum collected for recycling during one week can be shown as a bar graph or circle graph.

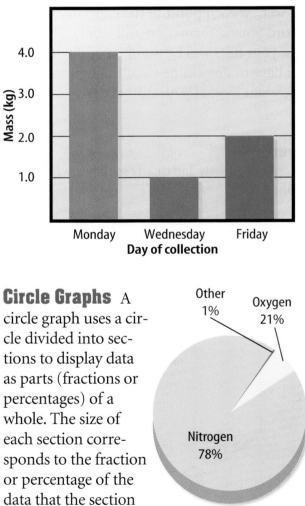

Circle Graphs A circle graph uses a circle divided into sections to display data as parts (fractions or percentages) of a whole. The size of each section corresponds to the fraction or percentage of the data that the section represents. So, the entire circle represents 100 percent, one-half represents 50 percent, one-fifth represents 20 percent, and so on.

Analyzing Results

To determine the meaning of your observations and investigation results, you will need to look for patterns in the data. You can organize your information in several of the ways that are discussed in this handbook. Then you must think critically to determine what the data mean. Scientists use several approaches when they analyze the data they have collected and recorded. Each approach is useful for identifying specific patterns in the data.

Forming Operational Definitions

An operational definition defines an object by showing how it functions, works, or behaves. Such definitions are written in terms of how an object works or how it can be used; that is, they describe its job or purpose.

For example, a ruler can be defined as a tool that measures the length of an object (how it can be used). A ruler also can be defined as something that contains a series of marks that can be used as a standard when measuring (how it works).

Classifying

Classifying is the process of sorting objects or events into groups based on common features. When classifying, first observe the objects or events to be classified. Then select one feature that is shared by some members in the group but not by all. Place those members that share that feature into a subgroup. You can classify members into smaller and smaller subgroups based on characteristics.

How might you classify a group of chemicals? You might first classify them by state of matter, putting solids, liquids, and gases into separate groups. Within each group, you

could then look for another common feature by which to further classify members of the group, such as color or how reactive they are.

Remember that when you classify, you are grouping objects or events for a purpose. For example, classifying chemicals can be the first step in organizing them for storage. Both at home and at school, poisonous or highly reactive chemicals should all be stored in a safe location where they are not easily accessible to small children or animals. Solids, liquids, and gases each have specific storage requirements that may include waterproof, airtight, or pressurized containers. Are the dangerous chemicals in your home stored in the right place? Keep your purpose in mind as you select the features to form groups and subgroups.

Figure 17
Color is one of many characteristics that are used to classify chemicals.

Comparing and Contrasting

Observations can be analyzed by noting the similarities and differences between two or more objects or events that you observe. When you look at objects or events to see how they are similar, you are comparing them. Contrasting is looking for differences in objects or events. The table below compares and contrasts the characteristics of two elements.

Elemental Characteristics		
Element	Aluminum	Gold
Color	silver	gold
Classification	metal	metal
Density (g/cm³)	2.7	19.3
Melting Point (°C)	660	1064

Recognizing Cause and Effect

Have you ever heard a loud pop right before the power went out and then suggested that an electric transformer probably blew out? If so, you have observed an effect and inferred a cause. The event is the effect, and the reason for the event is the cause.

When scientists are unsure of the cause of a certain event, they design controlled experiments to determine what caused it.

Interpreting Data

The word *interpret* means "to explain the meaning of something." Look at the problem originally being explored in an experiment and figure out what the data show. Identify the control group and the test group so you can see whether or not changes in the independent variable have had an effect. Look for differences in the dependent variable between the control and test groups.

These differences you observe can be qualitative or quantitative. You would be able to describe a qualitative difference using only words, whereas you would measure a quantitative difference and describe it using numbers. If there are differences, the independent variable that is being tested could have had an effect. If no differences are found between the control and test groups, the variable that is being tested apparently had no effect.

For example, suppose that three beakers each contain 100 mL of water. The beakers are placed on hot plates, and two of the hot plates are turned on, but the third is left off for a period of 5 min. Suppose you are then asked to describe any differences in the water in the three beakers. A qualitative difference might be the appearance of bubbles rising to the top in the water that is being heated but no rising bubbles in the unheated water. A quantitative difference might be a difference in the amount of water that is present in the beakers.

Inferring Scientists often make inferences based on their observations. An inference is an attempt to explain, or interpret, observations or to indicate what caused what you observed. An inference is a type of conclusion.

When making an inference, be certain to use accurate data and accurately described observations. Analyze all of the data that you've collected. Then, based on everything you know, explain or interpret what you've observed.

Drawing Conclusions

When scientists have analyzed the data they collected, they proceed to draw conclusions about what the data mean. These conclusions are sometimes stated using words similar to those found in the hypothesis formed earlier in the process.

Conclusions To analyze your data, you must review all of the observations and measurements that you made and recorded. Recheck all data for accuracy. After your data are rechecked and organized, you are almost ready to draw a conclusion such as "salt water boils at a higher temperature than freshwater."

Before you can draw a conclusion, however, you must determine whether the data allow you to come to a conclusion that supports a hypothesis. Sometimes that will be the case, other times it will not.

If your data do not support a hypothesis, it does not mean that the hypothesis is wrong. It means only that the results of the investigation did not support the hypothesis. Maybe the experiment needs to be redesigned, but very likely, some of the initial observations on which the hypothesis was based were incomplete or biased. Perhaps more observation or research is needed to refine the hypothesis.

Avoiding Bias Sometimes drawing a conclusion involves making judgments. When you make a judgment, you form an opinion about what your data mean. It is important to be honest and to avoid reaching a conclusion if no supporting evidence for it exists or if it was based on a small sample. It also is important not to allow any expectations of results to bias your judgments. If possible, it is a good idea to collect additional data. Scientists do this all the time.

For example, the *Hubble Space Telescope* was sent into space in April, 1990, to provide scientists with clearer views of the universe. *Hubble* is the size of a school bus and has a 2.4-m-diameter mirror. *Hubble* helped scientists answer questions about the planet Pluto.

For many years, scientists had only been able to hypothesize about the surface of the planet Pluto. *Hubble* has now provided pictures of Pluto's surface that show a rough texture with light and dark regions on it. This might be the best information about Pluto scientists will have until they are able to send a space probe to it.

Evaluating Others' Data and Conclusions

Sometimes scientists have to use data that they did not collect themselves, or they have to rely on observations and conclusions drawn by other researchers. In cases such as these, the data must be evaluated carefully.

How were the data obtained? How was the investigation done? Was it carried out properly? Has it been duplicated by other researchers? Were they able to follow the exact procedure? Did they come up with the same results? Look at the conclusion, as well. Would you reach the same conclusion from these results? Only when you have confidence in the data of others can you believe it is true and feel comfortable using it.

Communicating

The communication of ideas is an important part of the work of scientists. A discovery that is not reported will not advance the scientific community's understanding or knowledge. Communication among scientists also is important as a way of improving their investigations.

Scientists communicate in many ways, from writing articles in journals and magazines that explain their investigations and experiments, to announcing important discoveries on television and radio, to sharing ideas with colleagues on the Internet or presenting them as lectures.

People who study science rely on computers to record and store data and to analyze results from investigations. Whether you work in a laboratory or just need to write a lab report with tables, good computer skills are a necessity.

Using a Word Processor

Suppose your teacher has assigned a written report. After you've completed your research and decided how you want to write the information, you need to put all that information on paper. The easiest way to do this is with a word processing application on a computer.

A computer application that allows you to type your information, change it as many times as you need to, and then print it out so that it looks neat and clean is called a word processing application. You also can use this type of application to create tables and columns, add bullets or cartoon art to your page, include page numbers, and check your spelling.

Helpful Hints

- If you aren't sure how to do something using your word processing program, look in the help menu. You will find a list of topics there to click on for help. After you locate the help topic you need, just follow the step-by-step instructions you see on your screen.
- Just because you've spell checked your report doesn't mean that the spelling is perfect. The spell check feature can't catch misspelled words that look like other words. If you've accidentally typed *cold* instead of *gold*, the spell checker won't know the difference. Always reread your report to make sure you didn't miss any mistakes.

Figure 18
You can use computer programs to make graphs and tables.

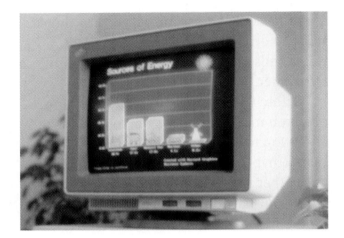

Using a Database

Imagine you're in the middle of a research project, busily gathering facts and information. You soon realize that it's becoming more difficult to organize and keep track of all the information. The tool to use to solve information overload is a database. Just as a file cabinet organizes paper records, a database organizes computer records. However, a database is more powerful than a simple file cabinet because at the click of a mouse, the contents can be reshuffled and reorganized. At computer-quick speeds, databases can sort information by any characteristics and filter data into multiple categories.

Helpful Hints

- Before setting up a database, take some time to learn the features of your database software by practicing with established database software.
- Periodically save your database as you enter data. That way, if something happens such as your computer malfunctions or the power goes off, you won't lose all of your work.

Doing a Database Search

When searching for information in a database, use the following search strategies to get the best results. These are the same search methods used for searching internet databases.

- Place the word *and* between two words in your search if you want the database to look for any entries that have both the words. For example, "gold *and* silver" would give you information that mentions both gold and silver.
- Place the word *or* between two words if you want the database to show entries that have at least one of the words. For example "gold *or* silver" would show you information that mentions either gold or silver.
- Place the word *not* between two words if you want the database to look for entries that have the first word but do not have the second word. For example, "gold *not* jewelry" would show you information that mentions gold but does not mention jewelry.

In summary, databases can be used to store large amounts of information about a particular subject. Databases allow biologists, Earth scientists, and physical scientists to search for information quickly and accurately.

Using an Electronic Spreadsheet

Your science fair experiment has produced lots of numbers. How do you keep track of all the data, and how can you easily work out all the calculations needed? You can use a computer program called a spreadsheet to record data that involve numbers. A spreadsheet is an electronic mathematical worksheet.

Type your data in rows and columns, just as they would look in a data table on a sheet of paper. A spreadsheet uses simple math to do data calculations. For example, you could add, subtract, divide, or multiply any of the values in the spreadsheet by another number. You also could set up a series of math steps you want to apply to the data. If you want to add 12 to all the numbers and then multiply all the numbers by 10, the computer does all the calculations for you in the spreadsheet. Below is an example of a spreadsheet that records test car data.

Helpful Hints

- Before you set up the spreadsheet, identify how you want to organize the data. Include any formulas you will need to use.
- Make sure you have entered the correct data into the correct rows and columns.
- You also can display your results in a graph. Pick the style of graph that best represents the data with which you are working.

Figure 19
A spreadsheet allows you to display large amounts of data and do calculations automatically.

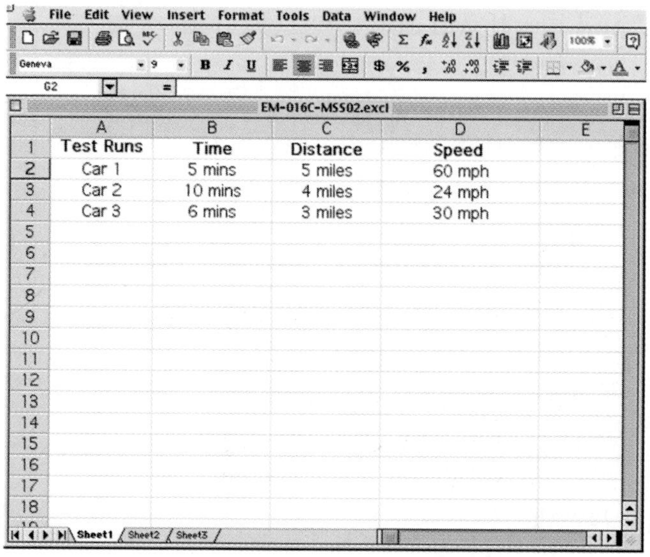

	A	B	C	D	E
1	Test Runs	Time	Distance	Speed	
2	Car 1	5 mins	5 miles	60 mph	
3	Car 2	10 mins	4 miles	24 mph	
4	Car 3	6 mins	3 miles	30 mph	

Using a Computerized Card Catalog

When you have a report or paper to research, you probably go to the library. To find the information you need in the library, you might have to use a computerized card catalog. This type of card catalog allows you to search for information by subject, by title, or by author. The computer then will display all the holdings the library has on the subject, title, or author requested.

A library's holdings can include books, magazines, databases, videos, and audio materials. When you have chosen something from this list, the computer will show whether an item is available and where in the library to find it.

Helpful Hints

- Remember that you can use the computer to search by subject, author, or title. If you know a book's author but not the title, you can search for all the books the library has by that author.
- When searching by subject, it's often most helpful to narrow your search by using specific search terms, such as *and, or,* and *not*. If you don't find enough sources this way, you can broaden your search.
- Pay attention to the type of materials found in your search. If you need a book, you can eliminate any videos or other resources that come up in your search.
- Knowing how your library is arranged can save you a lot of time. If you need help, the librarian will show you where certain types of materials are kept and how to find specific holdings.

Using Graphics Software

Are you having trouble finding that exact piece of art you're looking for? Do you have a picture in your mind of what you want but can't seem to find the right graphic to represent your ideas? To solve these problems, you can use graphics software. Graphics software allows you to create and change images and diagrams in almost unlimited ways. Typical uses for graphics software include arranging clip art, changing scanned images, and constructing pictures from scratch. Most graphics software applications work in similar ways. They use the same basic tools and functions. Once you master one graphics application, you can use other graphics applications.

Figure 20
Graphics software can use your data to draw bar graphs.

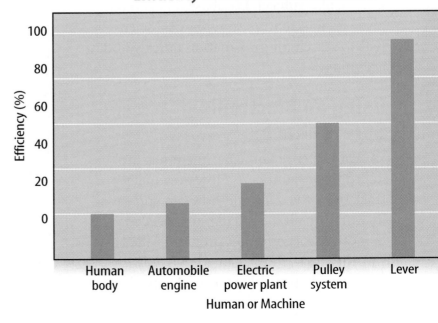

Efficiency of Humans and Machines

Figure 21
Graphics software can use your data to draw circle graphs.

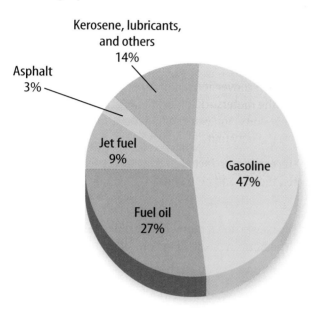

First, determine what important points you want to make in your presentation. Then, write an outline of what materials and types of media would best illustrate those points. Maybe you could start with an outline on an overhead projector, then show a video, followed by something from the Internet or a slide show accompanied by music or recorded voices. You might choose to use a presentation builder computer application that can combine all these elements into one presentation. Make sure the presentation is well constructed to make the most impact on the audience.

Figure 22
Multimedia presentations use many types of print and electronic materials.

Helpful Hints

- As with any method of drawing, the more you practice using the graphics software, the better your results will be.
- Start by using the software to manipulate existing drawings. Once you master this, making your own illustrations will be easier.
- Clip art is available on CD-ROMs and the Internet. With these resources, finding a piece of clip art to suit your purposes is simple.
- As you work on a drawing, save it often.

Developing Multimedia Presentations

It's your turn—you have to present your science report to the entire class. How do you do it? You can use many different sources of information to get the class excited about your presentation. Posters, videos, photographs, sound, computers, and the Internet can help show your ideas.

Helpful Hints

- Carefully consider what media will best communicate the point you are trying to make.
- Make sure you know how to use any equipment you will be using in your presentation.
- Practice the presentation several times.
- If possible, set up all of the equipment ahead of time. Make sure everything is working correctly.

Math Skill Handbook

Use this Math Skill Handbook to help solve problems you are given in this text. You might find it useful to review topics in this Math Skill Handbook first.

Converting Units

In science, quantities such as length, mass, and time sometimes are measured using different units. Suppose you want to know how many miles are in 12.7 km.

Conversion factors are used to change from one unit of measure to another. A conversion factor is a ratio that is equal to one. For example, there are 1,000 mL in 1 L, so 1,000 mL equals 1 L, or:

$$1{,}000 \text{ mL} = 1 \text{ L}$$

If both sides are divided by 1 L, this equation becomes:

$$\frac{1{,}000 \text{ mL}}{1 \text{ L}} = 1$$

The **ratio** on the left side of this equation is equal to 1 and is a conversion factor. You can make another conversion factor by dividing both sides of the top equation by 1,000 mL:

$$1 = \frac{1 \text{ L}}{1{,}000 \text{ mL}}$$

To **convert units,** you multiply by the appropriate conversion factor. For example, how many milliliters are in 1.255 L? To convert 1.255 L to milliliters, multiply 1.255 L by a conversion factor.

Use the **conversion factor** with new units (mL) in the numerator and the old units (L) in the denominator.

$$1.255 \text{ L} \times \frac{1{,}000 \text{ mL}}{1 \text{ L}} = 1{,}255 \text{ mL}$$

The unit L divides in this equation, just as if it were a number.

Example 1 There are 2.54 cm in 1 inch. If a meterstick has a length of 100 cm, how long is the meterstick in inches?

Step 1 Decide which conversion factor to use. You know the length of the meterstick in centimeters, so centimeters are the old units. You want to find the length in inches, so inch is the new unit.

Step 2 Form the conversion factor. Start with the relationship between the old and new units.

$$2.54 \text{ cm} = 1 \text{ inch}$$

Step 3 Form the conversion factor with the old unit (centimeter) on the bottom by dividing both sides by 2.54 cm.

$$1 = \frac{2.54 \text{ cm}}{2.54 \text{ cm}} = \frac{1 \text{ inch}}{2.54 \text{ cm}}$$

Step 4 Multiply the old measurement by the conversion factor.

$$100 \text{ cm} \times \frac{1 \text{ inch}}{2.54 \text{ cm}} = 39.37 \text{ inches}$$

The meterstick is 39.37 inches long.

Example 2 There are 365 days in one year. If a person is 14 years old, what is his or her age in days? (Ignore leap years)

Step 1 Decide which conversion factor to use. You want to convert years to days.

Step 2 Form the conversion factor. Start with the relation between the old and new units.

$$1 \text{ year} = 365 \text{ days}$$

Step 3 Form the conversion factor with the old unit (year) on the bottom by dividing both sides by 1 year.

$$1 = \frac{1 \text{ year}}{1 \text{ year}} = \frac{365 \text{ days}}{1 \text{ year}}$$

Step 4 Multiply the old measurement by the conversion factor:

$$14 \text{ years} \times \frac{365 \text{ days}}{1 \text{ year}} = 5{,}110 \text{ days}$$

The person's age is 5,110 days.

Practice Problem A book has a mass of 2.31 kg. If there are 1,000 g in 1 kg, what is the mass of the book in grams?

Using Fractions

A **fraction** is a number that compares a part to the whole. For example, in the fraction $\frac{2}{3}$, the 2 represents the part and the 3 represents the whole. In the fraction $\frac{2}{3}$, the top number, 2, is called the numerator. The bottom number, 3, is called the denominator.

Sometimes fractions are not written in their simplest form. To determine a fraction's **simplest form,** you must find the greatest common factor (GCF) of the numerator and denominator. The greatest common factor is the largest factor that is common to the numerator and denominator.

For example, because the number 3 divides into 12 and 30 evenly, it is a common factor of 12 and 30. However, because the number 6 is the largest number that evenly divides into 12 and 30, it is the **greatest common factor.**

After you find the greatest common factor, you can write a fraction in its simplest form. Divide both the numerator and the denominator by the greatest common factor. The number that results is the fraction in its **simplest form.**

Example Twelve of the 20 chemicals used in the science lab are in powder form. What fraction of the chemicals used in the lab are in powder form?

Step 1 Write the fraction.

$$\frac{\text{part}}{\text{whole}} = \frac{12}{20}$$

Step 2 To find the GCF of the numerator and denominator, list all of the factors of each number.

Factors of 12: 1, 2, 3, 4, 6, 12 (the numbers that divide evenly into 12)

Factors of 20: 1, 2, 4, 5, 10, 20 (the numbers that divide evenly into 20)

Step 3 List the common factors.

1, 2, 4.

Step 4 Choose the greatest factor in the list of common factors.

The GCF of 12 and 20 is 4.

Step 5 Divide the numerator and denominator by the GCF.

$$\frac{12 \div 4}{20 \div 4} = \frac{3}{5}$$

In the lab, $\frac{3}{5}$ of the chemicals are in powder form.

Practice Problem There are 90 rides at an amusement park. Of those rides, 66 have a height restriction. What fraction of the rides has a height restriction? Write the fraction in simplest form.

Calculating Ratios

A **ratio** is a comparison of two numbers by division.

Ratios can be written 3 to 5 or 3:5. Ratios also can be written as fractions, such as $\frac{3}{5}$. Ratios, like fractions, can be written in simplest form. Recall that a fraction is in **simplest form** when the greatest common factor (GCF) of the numerator and denominator is 1.

Example A chemical solution contains 40 g of salt and 64 g of baking soda. What is the ratio of salt to baking soda as a fraction in simplest form?

Step 1 Write the ratio as a fraction. $\dfrac{\text{salt}}{\text{baking soda}} = \dfrac{40}{64}$

Step 2 Express the fraction in simplest form. The GCF of 40 and 64 is 8.

$$\frac{40}{64} = \frac{40 \div 8}{64 \div 8} = \frac{5}{8}$$

The ratio of salt to baking soda in the solution is $\frac{5}{8}$.

Practice Problem Two metal rods measure 100 cm and 144 cm in length. What is the ratio of their lengths in simplest fraction form?

Using Decimals

A **decimal** is a fraction with a denominator of 10, 100, 1,000, or another power of 10. For example, 0.854 is the same as the fraction $\frac{854}{1,000}$.

In a decimal, the decimal point separates the ones place and the tenths place. For example, 0.27 means twenty-seven hundredths, or $\frac{27}{100}$, where 27 is the **number of units** out of 100 units. Any fraction can be written as a decimal using division.

Example Write $\frac{5}{8}$ as a decimal.

Step 1 Write a division problem with the numerator, 5, as the dividend and the denominator, 8, as the divisor. Write 5 as 5.000.

Step 2 Solve the problem.

```
      0.625
  8)5.000
      48
      ──
       20
       16
       ──
        40
        40
        ──
         0
```

Therefore, $\frac{5}{8} = 0.625$.

Practice Problem Write $\frac{19}{25}$ as a decimal.

Using Percentages

The word *percent* means "out of one hundred." A **percent** is a ratio that compares a number to 100. Suppose you read that 77 percent of Earth's surface is covered by water. That is the same as reading that the fraction of Earth's surface covered by water is $\frac{77}{100}$. To express a fraction as a percent, first find an equivalent decimal for the fraction. Then, multiply the decimal by 100 and add the percent symbol. For example, $\frac{1}{2} = 1 \div 2 = 0.5$. Then $0.5 \cdot 100 = 50 = 50\%$.

Example Express $\frac{13}{20}$ as a percent.

Step 1 Find the equivalent decimal for the fraction.

$$\begin{array}{r} 0.65 \\ 20\overline{)13.00} \\ \underline{120} \\ 100 \\ \underline{100} \\ 0 \end{array}$$

Step 2 Rewrite the fraction $\frac{13}{20}$ as 0.65.

Step 3 Multiply 0.65 by 100 and add the % sign.

$0.65 \cdot 100 = 65 = 65\%$

So, $\frac{13}{20} = 65\%$.

Practice Problem In one year, 73 of 365 days were rainy in one city. What percent of the days in that city were rainy?

Using Precision and Significant Digits

When you make a **measurement,** the value you record depends on the precision of the measuring instrument. When adding or subtracting numbers with different precision, the answer is rounded to the smallest number of decimal places of any number in the sum or difference. When multiplying or dividing, the answer is rounded to the smallest number of significant figures of any number being multiplied or divided. When counting the number of **significant figures,** all digits are counted except zeros at the end of a number with no decimal such as 2,500, and zeros at the beginning of a decimal such as 0.03020.

Example The lengths 5.28 and 5.2 are measured in meters. Find the sum of these lengths and report the sum using the least precise measurement.

Step 1 Find the sum.

5.28 m	2 digits after the decimal
+ 5.2 m	1 digit after the decimal
10.48 m	

Step 2 Round to one digit after the decimal because the least number of digits after the decimal of the numbers being added is 1.

The sum is 10.5 m.

Practice Problem Multiply the numbers in the example using the rule for multiplying and dividing. Report the answer with the correct number of significant figures.

Solving One-Step Equations

An **equation** is a statement that two things are equal. For example, $A = B$ is an equation that states that A is equal to B.

Sometimes one side of the equation will contain a **variable** whose value is not known. In the equation $3x = 12$, the variable is x.

The equation is solved when the variable is replaced with a value that makes both sides of the equation equal to each other. For example, the solution of the equation $3x = 12$ is $x = 4$. If the x is replaced with 4, then the equation becomes $3 \cdot 4 = 12$, or $12 = 12$.

To solve an equation such as $8x = 40$, divide both sides of the equation by the number that multiplies the variable.

$$8x = 40$$
$$\frac{8x}{8} = \frac{40}{8}$$
$$x = 5$$

You can check your answer by replacing the variable with your solution and seeing if both sides of the equation are the same.

$$8x = 8 \cdot 5 = 40$$

The left and right sides of the equation are the same, so $x = 5$ is the solution.

Sometimes an equation is written in this way: $a = bc$. This also is called a **formula.** The letters can be replaced by numbers, but the numbers must still make both sides of the equation the same.

Example 1 Solve the equation $10x = 35$.

Step 1 Find the solution by dividing each side of the equation by 10.

$$10x = 35 \qquad \frac{10x}{10} = \frac{35}{10} \qquad x = 3.5$$

Step 2 Check the solution.

$$10x = 35 \qquad 10 \times 3.5 = 35 \qquad 35 = 35$$

Both sides of the equation are equal, so $x = 3.5$ is the solution to the equation.

Example 2 In the formula $a = bc$, find the value of c if $a = 20$ and $b = 2$.

Step 1 Rearrange the formula so the unknown value is by itself on one side of the equation by dividing both sides by b.

$$a = bc$$
$$\frac{a}{b} = \frac{bc}{b}$$
$$\frac{a}{b} = c$$

Step 2 Replace the variables a and b with the values that are given.

$$\frac{a}{b} = c$$
$$\frac{20}{2} = c$$
$$10 = c$$

Step 3 Check the solution.

$$a = bc$$
$$20 = 2 \times 10$$
$$20 = 20$$

Both sides of the equation are equal, so $c = 10$ is the solution when $a = 20$ and $b = 2$.

Practice Problem In the formula $h = gd$, find the value of d if $g = 12.3$ and $h = 17.4$.

Using Proportions

A **proportion** is an equation that shows that two ratios are equivalent. The ratios $\frac{2}{4}$ and $\frac{5}{10}$ are equivalent, so they can be written as $\frac{2}{4} = \frac{5}{10}$. This equation is an example of a proportion.

When two ratios form a proportion, the **cross products** are equal. To find the cross products in the proportion $\frac{2}{4} = \frac{5}{10}$, multiply the 2 and the 10, and the 4 and the 5. Therefore $2 \cdot 10 = 4 \cdot 5$, **or** $20 = 20$.

Because you know that both proportions are equal, you can use cross products to find a missing term in a proportion. This is known as **solving the proportion.** Solving a proportion is similar to solving an equation.

Example The heights of a tree and a pole are proportional to the lengths of their shadows. The tree casts a shadow of 24 m at the same time that a 6-m pole casts a shadow of 4 m. What is the height of the tree?

Step 1 Write a proportion.

$$\frac{\text{height of tree}}{\text{height of pole}} = \frac{\text{length of tree's shadow}}{\text{length of pole's shadow}}$$

Step 2 Substitute the known values into the proportion. Let h represent the unknown value, the height of the tree.

$$\frac{h}{6} = \frac{24}{4}$$

Step 3 Find the cross products.

$$h \cdot 4 = 6 \cdot 24$$

Step 4 Simplify the equation.

$$4h = 144$$

Step 5 Divide each side by 4.

$$\frac{4h}{4} = \frac{144}{4}$$

$$h = 36$$

The height of the tree is 36 m.

Practice Problem The ratios of the weights of two objects on the Moon and on Earth are in proportion. A rock weighing 3 N on the Moon weighs 18 N on Earth. How much would a rock that weighs 5 N on the Moon weigh on Earth?

Math Skill Handbook

Statistics is the branch of mathematics that deals with collecting, analyzing, and presenting data. In statistics, there are three common ways to summarize the data with a single number—the mean, the median, and the mode.

The **mean** of a set of data is the arithmetic average. It is found by adding the numbers in the data set and dividing by the number of items in the set.

The **median** is the middle number in a set of data when the data are arranged in numerical order. If there were an even number of data points, the median would be the mean of the two middle numbers.

The **mode** of a set of data is the number or item that appears most often.

Another number that often is used to describe a set of data is the range. The **range** is the difference between the largest number and the smallest number in a set of data.

A **frequency table** shows how many times each piece of data occurs, usually in a survey. The frequency table below shows the results of a student survey on favorite color.

Color	Tally	Frequency
red	||||	4
blue	ⵌ	5
black	||	2
green	|||	3
purple	ⵌ ||	7
yellow	ⵌ |	6

Based on the frequency table data, which color is the favorite?

Example The speeds (in m/s) for a race car during five different time trials are 39, 37, 44, 36, and 44.

To find the mean:
Step 1 Find the sum of the numbers.

$$39 + 37 + 44 + 36 + 44 = 200$$

Step 2 Divide the sum by the number of items, which is 5.

$$200 \div 5 = 40$$

The mean measure is 40 m/s.

To find the median:
Step 1 Arrange the measures from least to greatest.

$$36, \ 37, \ \underline{39}, \ 44, \ 44$$

Step 2 Determine the middle measure.

The median measure is 39 m/s.

To find the mode:
Step 1 Group the numbers that are the same together.

$$44, 44, 36, 37, 39$$

Step 2 Determine the number that occurs most in the set.

$$\underline{44, 44}, 36, 37, 39$$

The mode measure is 44 m/s.

To find the range:
Step 1 Arrange the measures from largest to smallest.

$$44, 44, 39, 37, 36$$

Step 2 Determine the largest and smallest measures in the set.

$$\underline{44}, 44, 39, 37, \underline{36}$$

Step 3 Find the difference between the largest and smallest measures.

$$44 - 36 = 8$$

The range is 8 m/s.

Practice Problem Find the mean, median, mode, and range for the data set 8, 4, 12, 8, 11, 14, 16.

Safety in the Science Classroom

1. Always obtain your teacher's permission to begin an investigation.

2. Study the procedure. If you have questions, ask your teacher. Be sure you understand any safety symbols shown on the page.

3. Use the safety equipment provided for you. Goggles and a safety apron should be worn during most investigations.

4. Always slant test tubes away from yourself and others when heating them or adding substances to them.

5. Never eat or drink in the lab, and never use lab glassware as food or drink containers. Never inhale chemicals. Do not taste any substances or draw any material into a tube with your mouth.

6. Report any spill, accident, or injury, no matter how small, immediately to your teacher, then follow his or her instructions.

7. Know the location and proper use of the fire extinguisher, safety shower, fire blanket, first aid kit, and fire alarm.

8. Keep all materials away from open flames. Tie back long hair and tie down loose clothing.

9. If your clothing should catch fire, smother it with the fire blanket, or get under a safety shower. NEVER RUN.

10. If a fire should occur, turn off the gas then leave the room according to established procedures.

Follow these procedures as you clean up your work area

1. Turn off the water and gas. Disconnect electrical devices.

2. Clean all pieces of equipment and return all materials to their proper places.

3. Dispose of chemicals and other materials as directed by your teacher. Place broken glass and solid substances in the proper containers. Make sure never to discard materials in the sink.

4. Clean your work area. Wash your hands thoroughly after working in the laboratory.

First Aid	
Injury	**Safe Response ALWAYS NOTIFY YOUR TEACHER IMMEDIATELY**
Burns	Apply cold water.
Cuts and Bruises	Stop any bleeding by applying direct pressure. Cover cuts with a clean dressing. Apply ice packs or cold compresses to bruises.
Fainting	Leave the person lying down. Loosen any tight clothing and keep crowds away.
Foreign Matter in Eye	Flush with plenty of water. Use eyewash bottle or fountain.
Poisoning	Note the suspected poisoning agent.
Any Spills on Skin	Flush with large amounts of water or use safety shower.

REFERENCE HANDBOOK B

PERIODIC TABLE OF THE ELEMENTS

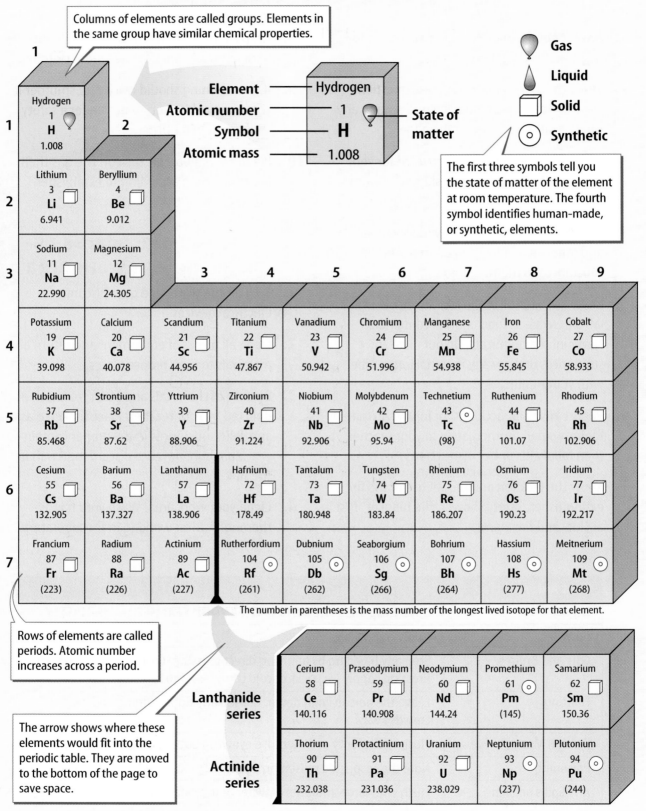

Columns of elements are called groups. Elements in the same group have similar chemical properties.

Gas
Liquid
Solid
Synthetic

Element — Hydrogen
Atomic number — 1
Symbol — H
Atomic mass — 1.008

State of matter

The first three symbols tell you the state of matter of the element at room temperature. The fourth symbol identifies human-made, or synthetic, elements.

1	2	3	4	5	6	7	8	9
1 Hydrogen 1 **H** 1.008								
2 Lithium 3 **Li** 6.941	Beryllium 4 **Be** 9.012							
3 Sodium 11 **Na** 22.990	Magnesium 12 **Mg** 24.305							
4 Potassium 19 **K** 39.098	Calcium 20 **Ca** 40.078	Scandium 21 **Sc** 44.956	Titanium 22 **Ti** 47.867	Vanadium 23 **V** 50.942	Chromium 24 **Cr** 51.996	Manganese 25 **Mn** 54.938	Iron 26 **Fe** 55.845	Cobalt 27 **Co** 58.933
5 Rubidium 37 **Rb** 85.468	Strontium 38 **Sr** 87.62	Yttrium 39 **Y** 88.906	Zirconium 40 **Zr** 91.224	Niobium 41 **Nb** 92.906	Molybdenum 42 **Mo** 95.94	Technetium 43 **Tc** (98)	Ruthenium 44 **Ru** 101.07	Rhodium 45 **Rh** 102.906
6 Cesium 55 **Cs** 132.905	Barium 56 **Ba** 137.327	Lanthanum 57 **La** 138.906	Hafnium 72 **Hf** 178.49	Tantalum 73 **Ta** 180.948	Tungsten 74 **W** 183.84	Rhenium 75 **Re** 186.207	Osmium 76 **Os** 190.23	Iridium 77 **Ir** 192.217
7 Francium 87 **Fr** (223)	Radium 88 **Ra** (226)	Actinium 89 **Ac** (227)	Rutherfordium 104 **Rf** (261)	Dubnium 105 **Db** (262)	Seaborgium 106 **Sg** (266)	Bohrium 107 **Bh** (264)	Hassium 108 **Hs** (277)	Meitnerium 109 **Mt** (268)

The number in parentheses is the mass number of the longest lived isotope for that element.

Rows of elements are called periods. Atomic number increases across a period.

The arrow shows where these elements would fit into the periodic table. They are moved to the bottom of the page to save space.

Lanthanide series	Cerium 58 **Ce** 140.116	Praseodymium 59 **Pr** 140.908	Neodymium 60 **Nd** 144.24	Promethium 61 **Pm** (145)	Samarium 62 **Sm** 150.36
Actinide series	Thorium 90 **Th** 232.038	Protactinium 91 **Pa** 231.036	Uranium 92 **U** 238.029	Neptunium 93 **Np** (237)	Plutonium 94 **Pu** (244)

Reference Handbook

Metal

Metalloid

Nonmetal

Recently discovered

The color of an element's block tells you if the element is a metal, nonmetal, metalloid, or has been discovered so recently that more study is needed.

SCIENCE *Online*

Visit the Glencoe Science Web site at **science.glencoe.com** for updates to the periodic table.

Reference Handbook

10	11	12	13	14	15	16	17	18
								Helium 2 He 4.003
			Boron 5 B 10.811	Carbon 6 C 12.011	Nitrogen 7 N 14.007	Oxygen 8 O 15.999	Fluorine 9 F 18.998	Neon 10 Ne 20.180
			Aluminum 13 Al 26.982	Silicon 14 Si 28.086	Phosphorus 15 P 30.974	Sulfur 16 S 32.065	Chlorine 17 Cl 35.453	Argon 18 Ar 39.948
Nickel 28 Ni 58.693	Copper 29 Cu 63.546	Zinc 30 Zn 65.39	Gallium 31 Ga 69.723	Germanium 32 Ge 72.64	Arsenic 33 As 74.922	Selenium 34 Se 78.96	Bromine 35 Br 79.904	Krypton 36 Kr 83.80
Palladium 46 Pd 106.42	Silver 47 Ag 107.868	Cadmium 48 Cd 112.411	Indium 49 In 114.818	Tin 50 Sn 118.710	Antimony 51 Sb 121.760	Tellurium 52 Te 127.60	Iodine 53 I 126.904	Xenon 54 Xe 131.293
Platinum 78 Pt 195.078	Gold 79 Au 196.967	Mercury 80 Hg 200.59	Thallium 81 Tl 204.383	Lead 82 Pb 207.2	Bismuth 83 Bi 208.980	Polonium 84 Po (209)	Astatine 85 At (210)	Radon 86 Rn (222)
Ununnilium * 110 Uun (281)	Unununium * 111 Uuu (272)	Ununbium * 112 Uub (285)		Ununquadium * 114 Uuq (289)		Ununhexium * 116 Uuh (289)		Ununoctium * 118 Uuo (293)

* Names not officially assigned. Discovery of elements 114, 116, and 118 recently reported. Further information not yet available.

Europium 63 Eu 151.964	Gadolinium 64 Gd 157.25	Terbium 65 Tb 158.925	Dysprosium 66 Dy 162.50	Holmium 67 Ho 164.930	Erbium 68 Er 167.259	Thulium 69 Tm 168.934	Ytterbium 70 Yb 173.04	Lutetium 71 Lu 174.967
Americium 95 Am (243)	Curium 96 Cm (247)	Berkelium 97 Bk (247)	Californium 98 Cf (251)	Einsteinium 99 Es (252)	Fermium 100 Fm (257)	Mendelevium 101 Md (258)	Nobelium 102 No (259)	Lawrencium 103 Lr (262)

REFERENCE HANDBOOK C

SI—Metric/English, English/Metric Conversions

	When you want to convert:	To:	Multiply by:
Length	inches	centimeters	2.54
	centimeters	inches	0.39
	yards	meters	0.91
	meters	yards	1.09
	miles	kilometers	1.61
	kilometers	miles	0.62
Mass and Weight*	ounces	grams	28.35
	grams	ounces	0.04
	pounds	kilograms	0.45
	kilograms	pounds	2.2
	tons (short)	tonnes (metric tons)	0.91
	tonnes (metric tons)	tons (short)	1.10
	pounds	newtons	4.45
	newtons	pounds	0.22
Volume	cubic inches	cubic centimeters	16.39
	cubic centimeters	cubic inches	0.06
	liters	quarts	1.06
	quarts	liters	0.95
	gallons	liters	3.78
Area	square inches	square centimeters	6.45
	square centimeters	square inches	0.16
	square yards	square meters	0.83
	square meters	square yards	1.19
	square miles	square kilometers	2.59
	square kilometers	square miles	0.39
	hectares	acres	2.47
	acres	hectares	0.40
Temperature	To convert °Celsius to °Fahrenheit		$°C \times 9/5 + 32$
	To convert °Fahrenheit to °Celsius		$5/9 (°F - 32)$

*Weight is measured in standard Earth gravity.

This glossary defines each key term that appears in bold type in the text. It also shows the chapter, section, and page number where you can find the words used.

A

acceleration: rate of change of velocity; can cause an object to speed up, slow down, or turn, and can be calculated by dividing the change in speed by the given time. (Chap. 1, Sec. 2, p. 14)

alternative resources: any renewable and inexhaustible sources of energy to generate electricity, including solar energy, wind, and geothermal energy. (Chap. 5, Sec. 3, p. 143)

Archimedes' principle: states that the buoyant force on an object equals the weight of the fluid displaced by the object. (Chap. 3, Sec. 2, p. 77)

B

balanced forces: two or more equal forces whose effects cancel each other out and do not change the motion of an object. (Chap. 2, Sec. 1, p. 37)

Bernoulli's principle: states that when the velocity of a fluid increases, the pressure exerted by the fluid decreases. (Chap. 3, Sec. 3, p. 85)

buoyant force: upward force exerted by a fluid on any object placed in the fluid. (Chap. 3, Sec. 2, p. 74)

C

chemical energy: energy that is stored in chemicals. (Chap. 5, Sec. 1, p. 129)

compound machine: device made up of a combination of two or more simple machines. (Chap. 4, Sec. 3, p. 109)

conduction: transfer of heat by direct contact; occurs when particles of one substance collide with particles of another substance, transferring kinetic energy. (Chap. 6, Sec. 2, p. 163)

conductor: any material that transfers heat easily. (Chap. 6, Sec. 2, p. 165)

convection: transfer of thermal energy by the movement of heated molecules from one place to another in a gas or liquid. (Chap. 6, Sec. 2, p. 164)

D

density: mass of an object divided by its volume. (Chap. 3, Sec. 2, p. 78)

E

efficiency: ability of a machine to convert input work to output work. (Chap. 4, Sec. 2, p. 107)

electrical energy: energy carried by electric current that comes out of batteries and wall sockets, is generated at large power plants, and is readily transformed into other types of energy. (Chap. 5, Sec. 1, p. 130)

energy: the ability to cause change. (Chap. 5, Sec. 1, p. 126)

engine: device that converts thermal energy into mechanical energy. (Chap. 6, Sec. 3, p. 169)

F

fluid: any substance that has no definite shape and can flow. (Chap. 3, Sec. 1, p. 69)

English Glossary

force: a push or a pull. (Chap. 2, Sec. 1, p. 36)

friction: rubbing force that acts against the motion between two touching surfaces and always slows an object down. (Chap. 2, Sec. 1, p. 38)

G

generator: device that transforms kinetic energy into electrical energy. (Chap. 5, Sec. 2, p. 136)

H

heat: thermal energy transferred from a warmer object to a cooler object. (Chap. 6, Sec. 2, p. 162)

hydraulic system: uses a fluid to increase an applied force. (Chap. 3, Sec. 3, p. 83)

I

inclined plane: simple machine with a flat, sloped surface, or ramp, that makes it easier to lift a heavy load by using less force over a greater distance. (Chap. 4, Sec. 3, p. 109)

inertia: tendency of an object to resist a change in its motion. (Chap. 1, Sec. 3, p. 19)

input force: force exerted on a machine; also called effort force. (Chap. 4, Sec. 2, p. 104)

instantaneous speed: the speed of an object at one instant of time. (Chap. 1, Sec. 1, p. 11)

internal combustion engine: engine in which fuel is burned in a combustion chamber inside the engine. (Chap. 6, Sec. 3, p. 170)

K

kinetic energy: energy an object has due to its motion. (Chap. 5, Sec. 1, p. 127)

L

law of conservation of energy: states that energy can change its form but it is never created or destroyed. (Chap. 5, Sec. 2, p. 132)

law of conservation of momentum: states that the total momentum of objects that collide with each other doesn't change. (Chap. 1, Sec. 3, p. 21)

lever: simple machine consisting of a rigid rod or plank that pivots or rotates about a fulcrum. (Chap. 4, Sec. 3, p. 112)

M

mass: amount of matter in an object. (Chap. 1, Sec. 3, p. 19)

mechanical advantage: number of times the input force is multiplied by a machine; can be calculated by dividing the output force by the input force. (Chap. 4, Sec. 2, p. 105)

momentum: a measure of how difficult it is to stop a moving object; a product of mass and velocity. (Chap. 1, Sec. 3, p. 20)

N

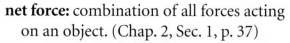

net force: combination of all forces acting on an object. (Chap. 2, Sec. 1, p. 37)

Newton's first law of motion: states that objects at rest will remain at rest or move with a constant velocity unless a force is applied. (Chap. 2, Sec. 1, p. 38)

Newton's second law of motion: states that an object acted upon by a net force will accelerate in the direction of the force. (Chap. 2, Sec. 2, p. 42)

Newton's third law of motion: states that forces always act in equal but opposite pairs. (Chap. 2, Sec. 3, p. 49)

nonrenewable resources: any energy sources that eventually will run out, such as coal and oil. (Chap. 5, Sec. 3, p. 140)

nuclear energy: energy stored in atomic nuclei that can be transformed into other forms of energy by very complex power plants. (Chap. 5, Sec. 1, p. 130)

O

output force: force exerted by a machine to overcome some resistance; also called resistance force. (Chap. 4, Sec. 2, p. 104)

P

Pascal's principle: states that the pressure applied at any point to a confined fluid is transmitted equally throughout the fluid. (Chap. 3, Sec. 3, p. 83)

photovoltaic: device that transforms radiant energy directly into electrical energy. (Chap. 5, Sec. 3, p. 144)

potential energy: energy stored in an object due to its position. (Chap. 5, Sec. 1, p. 128)

power: rate at which work is done. (Chap. 4, Sec. 1, p. 101)

pressure: amount of force applied per unit area on an object's surface; SI unit is the Pascal (Pa). (Chap. 3, Sec. 1, p. 66)

pulley: simple machine made from a grooved wheel with a rope or chain wrapped around the groove. (Chap. 4, Sec. 3, p. 114)

R

radiant energy: energy of light. (Chap. 5, Sec. 1, p. 129)

radiation: transfer of thermal energy by electromagnetic waves. (Chap. 6, Sec. 2, p. 163)

renewable resources: any energy sources that are replenished continually. (Chap. 5, Sec. 3, p. 142)

S

screw: simple machine made from an inclined plane wrapped around a cylinder. (Chap. 4, Sec. 3, p. 111)

simple machine: device that has only one movement; an inclined plane, lever, wheel and axle, and pulley. (Chap. 4, Sec. 3, p. 109)

specific heat: amount of energy necessary to raise the temperature of 1 kg of a substance by 1°C. (Chap. 6, Sec. 2, p. 166)

speed: rate of change of position, which can be calculated by dividing the distance traveled by the time it takes to travel that distance. (Chap. 1, Sec. 1, p. 10)

T

temperature: average value of the kinetic energy of the particles in a substance; can be measured using Fahrenheit, Celsius, and Kelvin scales. (Chap 6, Sec. 1, p. 158)

English Glossary

thermal energy: energy that all objects have; increases as the object's temperature increases. (Chap. 5, Sec. 1, p. 128)

thermal energy: total value of the kinetic and potential energy of a group of molecules. (Chap. 6, Sec. 1, p. 161)

thermal pollution: the increase in the temperature of a body of water caused by adding warmer water. (Chap. 6, Sec. 2, p. 167)

turbine: set of steam-powered fan blades that spins a generator at a power plant. (Chap. 5, Sec. 2, p. 136)

U

unbalanced forces: two or more unequal forces acting on an object, causing the object to accelerate. (Chap. 2, Sec. 1, p. 37)

V

velocity: speed and direction of a moving object. (Chap. 1, Sec. 1, p. 13)

W

wedge: simple machine consisting of an inclined plane that moves; can have one or two sloping sides. (Chap. 4, Sec. 3, p. 110)

weight: the gravitational force between you and Earth. (Chap. 2, Sec. 2, p. 43)

wheel and axle: simple machine made from two different-sized, circular objects that are attached and rotate together. (Chap. 4, Sec. 3, p. 112)

work: when a force exerted on an object causes that object to move some distance; is equal to force times distance. (Chap. 4, Sec. 1, p. 98)

Este glosario define cada término clave que aparece en negrillas en el texto. También muestra el capítulo, la sección y el número de página en donde se usa dicho término.

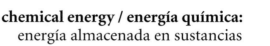

acceleration / aceleración: tasa de cambio en la velocidad; gracias a ella, un objeto puede acelerar, decelerar o girar; se puede calcular dividiendo el cambio en rapidez entre el tiempo dado. (Cap. 1, Sec. 2, pág. 14)

alternative resources / recursos alternos: toda fuente de energía, tanto renovable como inagotable, que se utiliza para generar electricidad; incluye la energía solar, la energía eólica y la energía geotérmica. (Cap. 5, Sec. 3, pág. 143)

Archimedes' principle / principio de Arquímedes: establece que la fuerza flotante en un cuerpo es igual al peso del fluido desplazado por dicho cuerpo. (Cap. 3, Sec. 2, pág. 77)

balanced forces / fuerzas equilibradas: dos o más fuerzas iguales cuyos efectos se anulan entre sí y no cambian el movimiento de un objeto. (Cap. 2, Sec. 1, pág. 37)

Bernoulli's principle / principio de Bernoulli: establece que al aumentar la velocidad de un fluido, la presión ejercida sobre el fluido disminuye. (Cap. 3, Sec. 3, pág. 85)

buoyant force / fuerza flotante: fuerza ascendente que un fluido ejerce sobre cualquier cuerpo dentro del fluido. (Cap. 3, Sec. 2, pág. 74)

chemical energy / energía química: energía almacenada en sustancias químicas. (Cap. 5, Sec. 1, pág. 129)

compound machine / máquina compuesta: dispositivo hecho de una combinación de máquinas simples. (Cap. 4, Sec. 3, pág. 109)

conduction / conducción: transferencia de energía por contacto directo; ocurre cuando las partículas de una sustancia chocan con las partículas de otra sustancia y transfieren energía cinética. (Cap. 6, Sec. 2, pág. 163)

conductor / conductor: cualquier material que transfiere energía fácilmente (Cap. 6, Sec. 2, pág. 165)

convection / convección: transferencia de energía térmica de un lugar a otro en un gas o un líquido debido al movimiento de moléculas calentadas. (Cap. 6, Sec. 2, pág. 164)

density / densidad: masa de un objeto dividida entre su volumen. (Cap. 3, Sec. 2, pág. 78)

efficiency / eficiencia: capacidad de una máquina de convertir el trabajo de entrada en trabajo de salida. (Cap. 4, Sec. 2, pág. 107)

electrical energy / energía eléctrica: energía transportada por la corriente eléctrica que sale de las pilas y de los enchufes de pared, se genera en centrales eléctricas grandes y se transforma fácilmente en otros tipos de energía. (Cap. 5, Sec. 1, pág. 130)

energy / energía: la capacidad de causar cambios. (Cap. 5, Sec. 1, pág. 126)

engine /motor: dispositivo que convierte la energía térmica en energía mecánica. (Cap. 6, Sec. 4, pág. 169)

F

fluid / fluido: cualquier sustancia que no posee forma definida y que puede fluir. (Cap. 3, Sec. 1, pág. 69)

force / fuerza: un empuje o un jalón. (Cap. 2, Sec. 1, pág. 36)

friction / fricción: fuerza frotadora que actúa contra el movimiento entre dos superficies en contacto y que siempre aminora la velocidad de un objeto. (Cap. 2, Sec. 1, pág. 38)

G

generator / generador: dispositivo que transforma la energía cinética en energía eléctrica. (Cap. 5, Sec. 2, pág. 136)

H

heat / calor: energía térmica transferida de un cuerpo más caliente a uno más frío. (Cap. 6, Sec. 2, pág. 162)

hydraulic system / sistema hidráulico: usa un fluido para aumentar la fuerza aplicada. (Cap. 3, Sec. 3, pág. 83)

I

inclined plane / plano inclinado: máquina simple con una superficie plana e inclinada, o rampa, que facilitar levantar cargas pesadas requiriendo menos fuerza a lo largo de una distancia mayor. (Cap. 4, Sec. 3, pág. 109)

inertia / inercia: tendencia que muestra un objeto de resistir cambios en su movimiento. (Cap. 1, Sec. 3, pág. 19)

input force / fuerza de entrada: fuerza que se ejerce sobre una máquina, también se conoce como fuerza de esfuerzo. (Cap. 4, Sec. 2, pág. 104)

instantaneous speed / rapidez instantánea: la rapidez de un cuerpo en un momento dado. (Cap. 1, Sec. 1, pág. 11)

internal combustion engine / motor de combustión interna: motor en que el combustible se quema en una cámara de combustión dentro del motor. (Cap. 6, Sec. 3, pág. 170)

kinetic energy / energía cinética: energía que tiene un cuerpo debido a su movimiento. (Cap. 5, Sec. 1, pág. 127)

L

law of conservation of energy / ley de conservación de la energía: establece que la energía puede transformarse pero nunca se crea ni se destruye. (Cap. 5, Sec. 2, pág. 132)

law of conservation of momentum / ley de conservación del momento: establece que el momento total de los cuerpos que chocan entre sí no cambia. (Cap. 1, Sec. 3, pág. 21)

lever / palanca: maquina simple que consta de una barra o tablón que gira alrededor de un fulcro. (Cap. 4, Sec. 3, pág. 112)

M

mass / masa: cantidad de materia que posee un cuerpo. (Cap. 1, Sec. 3, pág. 19)

mechanical advantage / ventaja mecánica: número de veces que una máquina multiplica una fuerza de entrada; se puede calcular dividiendo la fuerza de salida entre la fuerza de entrada. (Cap. 4, Sec. 2, pág. 105)

momentum / momento: medida del grado de dificultad que existe para detener un cuerpo en movimiento; el producto de la masa por la velocidad. (Cap. 1, Sec. 3, pág. 20)

N

net force / fuerza neta: combinación de todas las fuerzas que actúan sobre un objeto. (Cap. 2, Sec. 1, pág. 37)

Newton's first law of motion / primera ley del movimiento de Newton: establece que los objetos en reposo permanecerán en reposo o se moverán a una velocidad constante, a menos que se les aplique una fuerza. (Cap. 2, Sec. 1, pág. 38)

Newton's second law of motion / segunda ley del movimiento de Newton: establece que un objeto, al cual se le ha aplicado una fuerza neta, acelerará en la dirección de tal fuerza. (Cap. 2, Sec. 2, pág. 42)

Newton's third law of motion / tercera ley del movimiento de Newton: establece que las fuerzas siempre actúan en pares iguales pero opuestos. (Cap. 2, Sec. 3, pág. 49)

nonrenewable resources / recursos no renovables: toda fuente de energía que se agota a la larga, como el carbón y el petróleo. (Cap. 5, Sec. 3, pág. 140)

nuclear energy / energía nuclear: energía almacenada en los núcleos atómicos que se puede transformar en otras formas de energía en centrales eléctricas muy complejas. (Cap. 5, Sec. 1, pág. 130)

O

output force / fuerza de salida: fuerza ejercida por una máquina para sobreponer alguna resistencia; conocida también como fuerza de resistencia. (Cap. 4, Sec. 2, pág. 104)

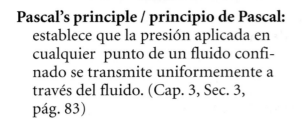

Pascal's principle / principio de Pascal: establece que la presión aplicada en cualquier punto de un fluido confinado se transmite uniformemente a través del fluido. (Cap. 3, Sec. 3, pág. 83)

photovoltaic / célula fotovoltaica: dispositivo que transforma la energía radiante directamente en energía eléctrica. (Cap. 5, Sec. 3, pág. 144)

potential energy / energía potencial: energía almacenada en un cuerpo debido a su posición. (Cap. 5, Sec. 1, pág. 128)

power / potencia: tasa a la cual se realiza trabajo. (Cap. 4, Sec. 1, pág. 101)

pressure / presión: cantidad de fuerza aplicada por unidad de área en la superficie de un objeto; la unidad SI es el Pascal (Pa). (Cap. 3, Sec. 1, pág. 66)

pulley / polea: máquina simple compuesta de una rueda acanalada con una cuerda o cadena enrollada alrededor de la parte acanalada. (Cap. 4, Sec. 3, pág. 114)

R

radiant energy / energía radiante: energía de la luz. (Cap. 5, Sec. 1, pág. 129)

radiation / radiación: transferencia de energía térmica por las ondas magnéticas. (Cap. 6, Sec. 2, pág. 163)

renewable resources / recursos renovables: toda fuente de energía que se regenera continuamente. (Cap. 5, Sec. 3, pág. 142)

S

screw / tornillo: máquina simple hecha de un plano inclinado enrollado alrededor de un cilindro. (Cap. 4, Sec. 3, pág. 111)

simple machine / máquina simple: dispositivo que sólo tiene un movimiento; plano inclinado, rueda y eje y polea. (Cap. 4, Sec. 3, pág. 109)

specific heat / calor específico: cantidad de energía necesaria para elevar 1°C la temperatura de un kg de una sustancia. (Cap. 6, Sec. 2, pág. 166)

speed / rapidez: tasa de cambio de posición, la cual se puede calcular dividiendo la distancia viajada entre el tiempo que se toma viajar tal distancia. (Cap. 1, Sec. 1, pág. 10)

T

temperature / temperatura: valor promedio de la energía cinética de las partículas en una sustancia; se puede medir usando una de las escalas Fahrenheit, Celsius o Kelvin. (Cap. 6, Sec. 1, pág. 158)

thermal energy / energía térmica: energía que tienen todos los cuerpos; aumenta conforme aumenta la temperatura del objeto. (Cap. 5, Sec. 1, pág. 128)

thermal energy / energía térmica: valor total de la energía cinética y la potencial de un grupo de partículas. (Cap. 6, Sec. 1, pág. 161)

thermal pollution/contaminación térmica: aumento en la temperatura de una masa de agua a raíz de agregar agua más caliente a la masa de agua. (Cap. 6, Sec. 2, pag. 167)

turbine / turbina: conjunto de álabes accionados a vapor que hace girar un generador en una central eléctrica. (Cap. 5, Sec. 2, pág. 136)

U

unbalanced forces / fuerzas desequilibradas: dos o más fuerzas desiguales que actúan sobre un objeto, haciendo que éste acelere. (Cap. 2, Sec. 1, pág. 37)

V

velocity / velocidad: rapidez y dirección de un cuerpo en movimiento. (Cap. 1, Sec. 1, pág. 13)

W

wedge / cuña: máquina simple que consta de un plano inclinado que se mueve; puede tener un o dos lados inclinados. (Cap. 4, Sec. 3, pág. 110)

weight / peso: la fuerza gravitatoria entre cualquier cuerpo y la Tierra. (Cap. 2, Sec. 2, pág. 43)

wheel and axle / rueda y eje: máquina simple compuesta de dos objetos circulares de distinto tamaño que están unidos y que giran juntos. (Cap. 4, Sec. 3, pág. 112)

work / trabajo: se hace trabajo cuando una fuerza ejercida sobre un objeto hace que el objeto se mueve cierta distancia; es igual a fuerza por distancia. (Cap. 4, Sec. 1, pág. 98)

The index for *Motion, Forces, and Energy* will help you locate major topics in the book quickly and easily. Each entry in the index is followed by the number of the pages on which the entry is discussed. A page number given in boldfaced type indicates the page on which that entry is defined. A page number given in italic type indicates a page on which the entry is used in an illustration or photograph. The abbreviation *act.* indicates a page on which the entry is used in an activity.

Index

Gravity, 43; air resistance and, 47, *47*

Index

Index

Index

Credits

Art Credits

Glencoe would like to acknowledge the artists and agencies who participated in illustrating this program: Absolute Science Illustration; Andrew Evansen; Argosy; Articulate Graphics; Craig Attebery represented by Frank & Jeff Lavaty; CHK America; Gagliano Graphics; Pedro Julio Gonzalez represented by Melissa Turk & The Artist Network; Robert Hynes represented by Mendola Ltd.; Morgan Cain & Associates; JTH Illustration; Laurie O'Keefe; Matthew Pippin represented by Beranbaum Artist's Representative; Precision Graphics; Publisher's Art; Rolin Graphics, Inc.; Wendy Smith represented by Melissa Turk & The Artist Network; Kevin Torline represented by Berendsen and Associates, Inc.; WILDlife ART; Phil Wilson represented by Cliff Knecht Artist Representative; Zoo Botanica.

Photo Credits

Abbreviation key: AA=Animals Animals; AH=Aaron Haupt; AMP=Amanita Pictures; BC=Bruce Coleman, Inc.; CB=CORBIS; DM=Doug Martin; DRK=DRK Photo; ES=Earth Scenes; FP=Fundamental Photographs; GH=Grant Heilman Photography; IC=Icon Images; KS=KS Studios; LA=Liaison Agency; MB=Mark Burnett; MM=Matt Meadows; PE=PhotoEdit; PD=PhotoDisc; PQ=PictureQuest; PR=Photo Researchers; SB=Stock Boston; TSA=Tom Stack & Associates; TSM=The Stock Market; VU=Visuals Unlimited.

Cover Gunter Marx Photography/CB; **vi** D.R. & T.L. Schrichte/Stone; **vii** Runk/Schoenberger from GH; **viii** First Image; **1** Globus Brothers Studios, NYC; **2** (t)Jeremy Woodhouse/PD, (b)Ted Spiegel/CB; **3** (t)William James Warren/CB, (b)CB; **5** Dominic Oldershaw; **6** Jeremy Woodhouse/PD; **6-7** Peter Griffith/Masterfile; **7** MB; **8** Telegraph Colour Library/FPG; **9** Geoff Butler; **12** Richard Hutchings; **15** Runk/Schoenberger from GH; **17** Mark Doolittle/Outside Images/PQ; **18** Rick Graves/Stone; **19** (l)TSM, (r)Will Hart/PE; **21** (t)Richard Megna/FP, (bl)Jodi Jacobson/Peter Arnold, Inc., (br)Jules Frazier/PD; **22** MB; **23** Slim Films; **24** Robert Brenner/PE; **25** Laura Sifferlin; **26** (t)Richard Olivier/CB, (b)IC; **27** IC; **29** Alexis Duclos/LA; **30** Tom & DeeAnn McCarthy/TSM; **31** (t)Rudi Von Briel/PE, (bl)AFP/CB, (br)PD; **34** Russell D. Curtis/PR; **34-35** Fujifotos/The Image Works; **35** Richard Hutchings; **36** (l)Globus Brothers Studios, NYC, (r)SB; **37** Bob Daemmrich; **38** (t)Beth Wald/Adventure Photo, (b)David Madison; **39** Rhoda Sidney/SB/PQ; **41** (l)Myrleen Cate/PE, (r)David Young-Wolff/PE; **42** Bob Daemmrich; **44** (t)Stone, (b)Myrleen Cate/PE; **46** David Madison; **48** (t)Tom Sanders/Adventure Photo, (b)Richard Fuller/David Madison Sports Images; **49** Mary M. Steinbacher/PE; **50** (t)Betty Sederquist/VU, (b)Jim Cummins/FPG; **51** (tl)Denis Boulanger/Allsport, (tr)Donald Miralle/Allsport, (b)Tony Freeman/PE/PQ; **52** (t)David Madison, (b)NASA; **54** NASA; **55** Richard Hutchings; **56 57** First Image; **58** Didier Charre/The Image Bank; **59** Tom Wright/CB; **60** (t)William R. Sallaz/Duomo, (c)Bob Daemmrich, (b)First Image; **61** (t)Philip Bailey/TSM, (c)Romilly Lockyer/The Image Bank, (b)Tony Freeman/PE; **64** Jose Azel/Aurora/PQ; **64-65** Martin Rogers/Stone; **65** MM; **66** David Young-Wolff/PE; **68** Runk/Schoenberger from GH; **69** Dominic

Oldershaw; **70** (t)MM, (b)Tom Pantages; **72** (bkgd)Janet Dell Russell Johnson, (t)Bobby Model/National Geographic Image Collection, (cl)Richard Nowitz/National Geographic Image Collection, (cr)George Grall/National Geographic Image Collection, (bl)Ralph White/CB, (br)CB; **74** Ryan McVay/PD; **75** CB; **76** (t)MM, (b)Vince Streano/Stone; **77 79** MM; **81** John Evans; **83** KS; **84** Dominic Oldershaw; **87** (t)Michael Collier/SB, (c)George Hall/CB, (b)Dean Conger/CB; **88** (t)Runk/Schoenberger from GH, (b)Steve McCutcheon/VU; **90** Ray Fairall/AP/Wide World Photos; **91** Courtesy Joan A. Suggs-Cooper; **92** (t)Keren Su/CB, (c)James A. Sugar/Black Star Publishing/PQ, (b)Kevin R. Morris/CB; **93** (l)D.R. & T.L. Schrichte/Stone, (r)CB; **94** Thomas Hovland from GH; **96** Sandia National Laboratories; **96-97** Dan Habib/Impact Visuals/PQ; **97** MB; **98** Mary Kate Denny; **99** (tl tr)Richard Hutchings, (b)Tony Freeman/PE; **106** (l)Frank Siteman/SB, (r)David Young-Wolff/PE; **109** Duomo; **110** Robert Brenner/PE; **111** (t)Tom McHugh/PR; **111** (b)AMP; **112** AMP; **113** (t)Dorling Kindersley, (bl br)Bob Daemmrich; **114** (l)Siegfried Layda/Stone, (r)Wernher Krutein/LA; **116** (t)Tony Freeman/PE, (b)Tony Page/Stone; **117** AH; **118** (l)Ed Kashi/CB, (r)Secci-Lecaque/Roussel-USCLAF/CNRI/Science Photo Library/PR; **119** (t)Keri Pickett, (b)James Balog/Contact; **120** (tl)Gabe Palmer/TSM, (tr)Ken Frick, (b)StudiOhio; **121** (l)Inc. Janeart/The Image Bank, (r)Ryan McVay/PD; **124** Charles Krebs/Stone; **124-125** Roger Ressmeyer/CB; **125** Bob Daemmrich; **126** (l c)file photo, (r)MB; **127** (t b)Bob Daemmrich, (c)Al Tielemans/Duomo; **128** KS; **129** (tl tr)Bob Daemmrich, (br)Andrew McClenaghan/Science Photo Library/PR; **130** MB/PR; **131** Lori Adamski Peek/Stone; **132** Richard Hutchings; **133** Ron Kimball/Ron Kimball Photography; **134** (tl)Judy Lutz, (tc tr bl)Stephen R. Wagner, (br)Lennart Nilsson; **136 138** KS; **144** (t)Dr. Jeremy Burgess/Science Photo Library/PR, (b)John Keating/PR; **145** Geothermal Education Office; **148** Carsand-Mosher; **149** Roger Ressmeyer/CB; **150** SuperStock; **152** (t)James Blank/FPG, (c)Robert Torres/Stone, (bl br)SuperStock; **153** (l)Lowell Georgia/CB, (r)Mark Richards/PE; **154** Reuters NewMedia/CB; **156** Archive Photos; **156-157** Dave Jacobs/Stone; **157** AH; **158** John Evans; **159** (t)Nancy P. Alexander/VU, (b)Morton & White; **161** Tom Stack/TSA; **162** DM; **163** MM; **164** Jeremy Hoare/PD; **165** Donnie Kamin/PE; **166** SuperStock; **167** Colin Raw/Stone; **168** AH; **170** (l)Barbara Stitzer/PE, (c)Doug Menuez/PD, (r)Addison Geary/SB; **171** Slim Films; **172** C. Squared Studios/PD; **174 175** Morton & White; **176-177** Chip Simons/FPG; **177** Joseph Sohm/CB; **178** (tl)James Holmes/Science Photo Library/PR, (tr)Jenny Hager/The Image Works, (b)Charles D. Winters/PR; **179** SuperStock; **184-185** PD; **186** file photo; **187** (t)Dan Feicht, (b)VU; **188** Jose Carrillo/PE; **189** (t)AH, (b)Michael J. Howell/Rainbow/PQ; **190** NASA; **191** (l)NASA, (r)Roger Ressmeyer/CB; **192** (t)NASA/Roger Ressmeyer/CB, (c b)NASA; **193** NASA; **194** Timothy Fuller; **198** Roger Ball/TSM; **200** (l)Geoff Butler, (r)Coco McCoy/Rainbow/PQ; **201** Dominic Oldershaw; **202** StudiOhio; **203** First Image; **205** MM; **208** Paul Barton/TSM; **211** Davis Barber/PE.

Acknowledgments

"Hurricane" by John Balaban. Reprinted by permission of the author.

PERIODIC TABLE OF THE ELEMENTS

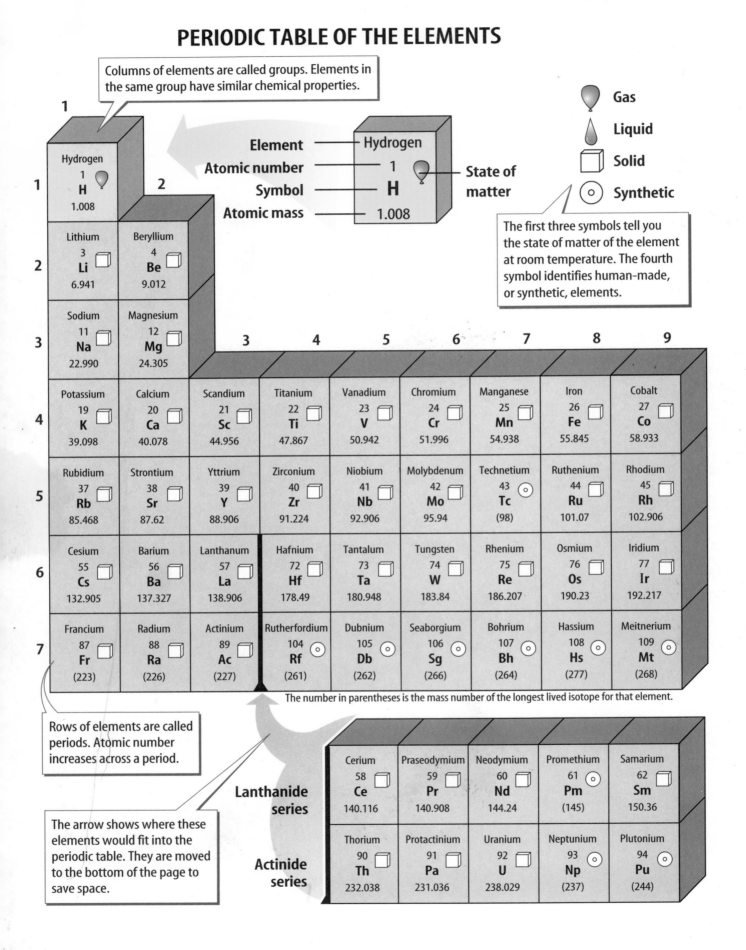

Columns of elements are called groups. Elements in the same group have similar chemical properties.

Gas

Liquid

Solid

Synthetic

Element — Hydrogen
Atomic number — 1
Symbol — H
Atomic mass — 1.008

State of matter

The first three symbols tell you the state of matter of the element at room temperature. The fourth symbol identifies human-made, or synthetic, elements.

1								
1 Hydrogen 1 **H** 1.008	**2**							
2 Lithium 3 **Li** 6.941	Beryllium 4 **Be** 9.012							
3 Sodium 11 **Na** 22.990	Magnesium 12 **Mg** 24.305	**3**	**4**	**5**	**6**	**7**	**8**	**9**
4 Potassium 19 **K** 39.098	Calcium 20 **Ca** 40.078	Scandium 21 **Sc** 44.956	Titanium 22 **Ti** 47.867	Vanadium 23 **V** 50.942	Chromium 24 **Cr** 51.996	Manganese 25 **Mn** 54.938	Iron 26 **Fe** 55.845	Cobalt 27 **Co** 58.933
5 Rubidium 37 **Rb** 85.468	Strontium 38 **Sr** 87.62	Yttrium 39 **Y** 88.906	Zirconium 40 **Zr** 91.224	Niobium 41 **Nb** 92.906	Molybdenum 42 **Mo** 95.94	Technetium 43 **Tc** (98)	Ruthenium 44 **Ru** 101.07	Rhodium 45 **Rh** 102.906
6 Cesium 55 **Cs** 132.905	Barium 56 **Ba** 137.327	Lanthanum 57 **La** 138.906	Hafnium 72 **Hf** 178.49	Tantalum 73 **Ta** 180.948	Tungsten 74 **W** 183.84	Rhenium 75 **Re** 186.207	Osmium 76 **Os** 190.23	Iridium 77 **Ir** 192.217
7 Francium 87 **Fr** (223)	Radium 88 **Ra** (226)	Actinium 89 **Ac** (227)	Rutherfordium 104 **Rf** (261)	Dubnium 105 **Db** (262)	Seaborgium 106 **Sg** (266)	Bohrium 107 **Bh** (264)	Hassium 108 **Hs** (277)	Meitnerium 109 **Mt** (268)

The number in parentheses is the mass number of the longest lived isotope for that element.

Rows of elements are called periods. Atomic number increases across a period.

The arrow shows where these elements would fit into the periodic table. They are moved to the bottom of the page to save space.

Lanthanide series	Cerium 58 **Ce** 140.116	Praseodymium 59 **Pr** 140.908	Neodymium 60 **Nd** 144.24	Promethium 61 **Pm** (145)	Samarium 62 **Sm** 150.36
Actinide series	Thorium 90 **Th** 232.038	Protactinium 91 **Pa** 231.036	Uranium 92 **U** 238.029	Neptunium 93 **Np** (237)	Plutonium 94 **Pu** (244)